RENEWALS 458-4574

DATE DUE

GAYLORD			PRINTED IN U.S.A

Portraits of Discovery

Also by George Greenstein

Frozen Star: Of Pulsars, Black Holes and the Fate of Stars
The Symbiotic Universe: Life and Mind in the Cosmos
The Quantum Challenge (with Arthur Zajonc)

Portraits of Discovery

Profiles in Scientific Genius

George Greenstein

John Wiley & Sons, Inc.

New York • Chichester • Weinheim • Brisbane • Singapore • Toronto

Library of Congress Cataloging-in-Publication Data
Greenstein, George.
    Portraits of discovery : profiles in scientific genius
George Greenstein.
        p.   cm.
    Includes bibliographical references and index.
    ISBN 0-471-19138-8 (cloth : alk. paper)
    1. Scientists—Biography. 2. Discoveries in science.   I. Title
Q141.G775   1997
509.2'2—dc21   97-6048
[B]

Printed in the United States of America

10 9 8 7 6 5 4 3 2 1

To Ilana

Contents

Prologue

The Other World

Not long ago a meteor fell to earth not far from where I live. The thing landed right on top of somebody's home: punched a hole in the roof, knocked over a lamp, and came to rest on the carpet in the front hall. News broadcasts that evening showed the owners standing around looking bemused while insurance adjusters went over the damage.

So . . . duly noted, and soon to be forgotten. But as the days passed, I found that I was not forgetting that meteor. My mind kept circling back to it—but from different angles than those reported in the news. I found myself thinking of it not as a stone that fell to earth, but as a silent voyager—an isolated, lonely chunk that had floated undisturbed through interplanetary voids for untold ages before being slammed into by this great, noisy planet of ours. For how long? Millions of years? Billions? You could tell by looking for the microscopic tracks of cosmic rays etched within it: the more tracks, the longer it had been up there. On other days I caught myself wondering if the thing had been spinning, tumbling gently end over end until the collision. Probably: everything else in the solar system rotated—planets, moons. Why this should be so was something of a question, but the facts were clear. And yet another question: How had it gotten up there in the first place? It takes intense pressure to form hard stone, a pressure difficult to come by in the empty depths of space. The meteor could only have been formed deep in the heart of some nameless extraterrestrial body, which must then have broken up, sending it off into space.

So I played with the meteor for several days, before eventually dropping the matter and moving on to other things. I don't want to give the wrong impression, though. This book is not about meteors. It is about scientists. The moral I want to draw from my story has nothing to do with the meteor, but plenty to do with me. The way I had thought about it other people might find strange—but it is utterly characteristic of the mind of the scientist. Anybody could see that stone lying on the carpet, the shattered lamp, and the hole in the roof. But the scientist is likely to peer not so much *at* the hole as *through* it—through it, and out into the Other World.

Scientists are convinced that there's more going on than meets the eye. They keep seeing clues to a world lying beyond that of daily reality. Luis Alvarez looked at the unusual chemical composition of a certain geologic stratum, and he saw traces of a prehistoric catastrophe, an ancient cataclysm in which the dinosaurs went extinct. Annie Jump Cannon looked at a rainbow, and at rainbows cast by the distant stars, and she found a regularity—a regularity that remained mysterious for years, until a graduate student named Cecilia Payne came along and explained it. Homi Bhabha looked at a cosmic ray particle, and what he saw was a tiny, rapidly moving clock. George Gamow looked at the Milky Way, and he found traces of the grandest story of them all, that of creation.

The Other World is hard to see. It does not lie about us so obviously as tables and chairs. You have to work to catch a glimpse of it. Martin Perl spots it in the computerized output of a particle detector the size of a house, sitting at the business end of a 2-mile-long linear accelerator. Ludwig Boltzmann saw it in his mind's eye as he gazed at pages upon pages of mathematics, and found within them a new insight into the nature of time. You might think that John Huchra sees it through a telescope on a mountaintop in Arizona, but I'm not so sure that's so. It's really not till he gets back home and goes over the data with Margaret Geller that the two catch glimpses of our address in the universe.

In this book I want to tell the stories of these people, and of others too. Some are famous—Feynman in America, Homi Bhabha in India. Others are not so widely known. But they should be: the stories are all good ones, worth telling.

I want to tell them in all their richness of detail. I want to do justice to the work these people did—and the only way to do this is to pay close attention to the science itself. I have no use for accounts of great discoveries that tell us all about the lives of the people involved, but that somehow manage to avoid mentioning much of what they actually did. For after all, the work is what these people's lives are *about*. It is what gives their lives meaning; what animates their days and consoles their nights. I want to pay to their efforts the same respect that they themselves bring to it.

Science passes its practitioners by. Most science textbooks hardly even mention the names of the people who did the important work. They present the material in isolation, disconnected from the story of its discovery. The work becomes anonymous. But I have always felt that this practice squeezes away much of the lifeblood of these discoveries. Perhaps this is why so many people regard science as being cold and sterile. Somehow, textbooks often end up conveying an image of science as nothing more than a gigantic list of facts—just a lot of complicated stuff to be memorized. But this has nothing to do with the actual practice of research, its warm, human vibrancy and passionate intensity. I want to convey something of this passion in these pages.

For in truth, there are wonderful stories to tell. I wish I had known George Gamow personally. He was rough, jovial, exuberant, and friendly; addicted to magic tricks, games, spoofs, elaborate parodies of his colleagues and their theories. He would spend hours constructing elaborate semiscientific gizmos out of plywood and string, festooned with legends in various languages. Ludwig Boltzmann had initially interested me because of the beauty of his work. But when I looked into him, I found more than I had expected: a man formidable, imposing, and severe in his professional life, a relentless battler, withering in his criticism of theories he opposed—but sweet, unassuming, and generous in his personal relations, a man who would weep openly at the railroad station if forced to leave his wife even briefly, and who was so crushed by mental illness that in the end he committed suicide. As for Richard Feynman, his books of reminiscences are the only autobiographies I have ever read that regularly sent me into paroxysms of laughter. The guy was a stitch—a nut. He was also a man whose research was breathtaking in its power and brilliance; and a man whose technical papers are written in a style so elegant, so perfect in its lucidity and refinement, as to remind me of nothing so much as the music of Mozart.

I set out to write of the scientific work of my subjects and of their lives—but in many cases I found I could not do so without venturing into other areas. Whether they wished it or not, the careers of many of these people were intimately bound up in political or social controversies. Three of my subjects—Annie Jump Cannon, Cecilia Payne, and Margaret Geller—are women, and much of what I have to tell about them concerns the status of women in science. It is a disgraceful story, one of which we should all be ashamed. Luis Alvarez was asked to testify in the hearings that led to the destruction of J. Robert Oppenheimer's career in government: after much soul-searching, he agreed to do so, a decision that tainted him in the eyes of many for the rest of his life. Most of Homi Bhabha's work was concerned with the development of science in India, a nation beset by problems so overwhelming as to make the practice of science incomparably more difficult than here. And Martin Perl's career has been intimately bound up with big science—a striking and relatively recent phenomenon; one that has transformed the very nature of the profession, and that has brought it into a strained, uneasy relationship with government and with the public upon which it depends.

There's a tension winding through the pages of this book. The scientist deals with the universal, the abstract, and the permanent—but the scientist's life is particular, individual, and unique. So much of a person's career is subject to the vagaries of chance, to stray circumstances, and to the luck of the draw. And yet, one way or another, out of this flux something solid emerges: something verifiable, and as true as we can make it. I am describing in this book the encounters of individual, contingent lives with the hard rock of external truth.

These people have been writing our address in the universe—and what they have found is that we are way out in the outskirts. We and our concerns are nowhere near the center of the action. The map of the universe that Margaret Geller and John Huchra are busy drawing encompasses regions so gigantic as to shrink our entire planet Earth into invisibility. At the same time, however, many of the most important constituents of the world, such as Martin Perl's particle, are inconceivably smaller than we. The cosmos has lasted for billions of years, while some things happen in a billionth of a second. It is abstract: the very terms in which we must come to grips with it are not the terms with which we live our lives. It does not care about us.

You'd think a lifetime of immersion in so inhuman a universe would make scientists downcast and gloomy. You'd think they would be overwhelmed by what they have found. But they are not. As a matter of fact, the truth is just the opposite. The pursuit of science is a liberating, invigorating experience. There is a great, windy sense of freedom to the whole enterprise. Someone once said that the wonderful thing about being a scientist is that it gives you something to be smart *about*. No matter how much you know, you do not know enough. No matter how creative you are, you are not being creative enough. No mater how far you have gone, the road winds on.

And the terrain it winds through is extraordinary. For in truth, the universe is a marvelous place, shot through with many wonders. It has turned out to be richer by far than we had any right to hope. Every advance in our understanding makes the world more astonishing, not less. Every discovery opens up ever deeper mysteries, ever more lovely vistas. The more we learn, the more we realize that the universe has been vastly underrated.

1

The Ladies of Observatory Hill

Annie Jump Cannon and
Cecilia Payne-Gaposchkin

One evening in 1919, a student at Cambridge University named Cecilia Payne attended a public lecture by a famous astronomer. By Payne's account her presence that night was something of an accident; her college had been allotted only four tickets, but at the last minute one of her friends was unable to go. I think it is no exaggeration to say that this accident changed her life. After the lecture "when I returned to my room," Cecilia wrote in her autobiography decades later, "I found that I could write down the lecture word for word. . . . For three nights, I think, I did not sleep. My world had been so shaken that I experienced something very like a nervous breakdown." She decided to become an astronomer.

The lecturer's name was Eddington: he will appear later in this story. Some time later, Payne encountered him again, and she told him of her wish. "I can see no *insuperable* objection," Eddington replied after a meditative pause. Cecilia asked him what to read. He mentioned several books. She had already read them. He mentioned several journals.

Payne was off in a flash, and she seems never to have slowed down. Not long after, she bicycled up to the observatory with a question in mind. She found someone astride the roof, conducting repairs. "I have come," she yelled up at him, "to ask why the Stark Effect is not observed in stellar spectra."

Cecilia Payne had always been interested in science. As a child she once performed an experimental determination of the efficacy of prayer. She divided her school examinations into two groups, praying for success in one while failing to mention the others. (She ended up doing better in the second group.) She must, in fact, have been a delightful child. On another occasion she wrote a tragedy in blank verse on a scientist who allowed himself to be corrupted by money, finally atoning by becoming a farmer. Nor had she neglected the arts: at the age of twelve she had given a piano recital, and had asked the teacher if she could say a few words beforehand. Permission granted, Cecilia told the audience that her piece was so beautiful it ought never to be played.

She had been born in 1900 and raised in rural England. In 1923 she left England to begin work for her graduate degree at Harvard. She already knew what she wanted to do there. She wanted to explain a Grand Fact.

"We have without doubt in the heavens a grand fact," wrote the director of the Vatican Observatory in 1868, "the fundamental distinction between the stars according to a small number of types." Stars came in categories. It is not particularly obvious to the naked eye that the stars are of different types. After all, a glance at the sky reveals them to look pretty much alike. They do lie in constellations—here a dipper, there a swan—but these patterns have no significance: stars scatter more or less randomly across the sky, and it is merely by accident that their arrangement occasionally reminds us of something we have seen before. Similarly, some stars are bright and some are dim, but apparent brightness depends on the random happenstance of distance: the bright ones may not be unusually luminous, but just rather close.

Even as Payne was unpacking her bags at Harvard, the true differences between stars were finally being codified in a monumental nine-volume study by the Harvard astronomer Annie Jump Cannon. Cannon's scheme of classification rests on a phenomenon invisible to the naked eye. Hanging in her home on the Observatory grounds was one of those lovely old chandeliers composed of glass prisms. As these prisms caught the sunlight, they cast a multitude of tiny rainbows upon her walls. White light, when broken into its constituent colors, forms a spectrum: a rainbow is the spectrum of our Sun. Nighttimes at the telescope, Cannon and other astronomers passed not sunlight but starlight through just such a prism, and they recorded on photographs the spectra, not of the Sun, but of the incomparably more distant stars.

These spectra turned out to differ from that of the Sun—if the distant stars did possess planets, rainbows upon them would differ from those on the Earth. And the spectra differed from one another. But not completely so: stellar spectra were found to come in classes. It was the identification of these classes, and the ordering according to this scheme of a great number of stars, that was Annie Jump Cannon's great contribution to astronomy.

She had been born in 1863 in Delaware. Her father, a shipbuilder and lieutenant governor of the state, had cast the deciding vote against secession when the Civil War broke out. As a child she was fascinated by the nighttime sky, a love she shared with her mother, who had once taken a course in astronomy. The two would conduct observations from a makeshift observatory in the attic of their large house, taking notes by candlelight. It was rare for young women to receive an advanced education in those days, but Annie's parents sent her to Wellesley College in Massachusetts. There she studied physics. But upon graduating, she returned home, and for the next decade she made not the slightest effort to continue professionally.

It was the death of her mother, to whom she was very close, that seems to have propelled Annie Cannon into her career. She returned to Wellesley for further study, and then moved on to Radcliffe. In 1894 she was hired by the director of the Harvard Observatory, and set to work attempting to make some sense of its vast collection of stellar spectra. It was a project that would occupy her for her entire life.

Every account I have ever read of Annie Jump Cannon mentions her serene, affectionate, and engaging nature. In her obituary of Cannon (she died in 1941), Payne writes that she was the happiest person Payne had ever known. Cannon seems to have had the knack of forming warm and lasting friendships with everyone. She carried on a lifelong and extensive correspondence with family, childhood friends, college classmates, amateur and professional astronomers. Although she never married, she loved children and would regularly host parties for them in Star Cottage, her home on the observatory grounds. A photograph of her shows an aged lady dandling a child on her knee, surrounded by kids in party hats. She herself was an avid photographer, and once produced a booklet of her photographs to be used by a camera company as a souvenir at the 1893 World's Fair.

In many ways the Harvard Observatory in those days must have been a delightful place. Situated on Observatory Hill, a mile or so up the road from the main campus, it was both physically and administratively somewhat separate from Harvard and Radcliffe. When Payne arrived, only Harlow Shapley, the director, carried a university appointment. Many astronomers lived on the observatory grounds and hardly ever mingled with the university faculty: at one point Payne wrote that in her seven years there, her acquaintances had been limited to observatory staff and one physics professor. Staff were regularly invited—and expected—to attend parties at the director's residence: there Shapley, jauntily set out in derby hat, would lead square dances and Virginia reels. One astronomer kept in his office a small box bearing the label "flashlight bulbs, probably burnt out."

When Cannon began her work, the study of stellar spectra was in a state that can only be described as chaotic. By the turn of the century, more than twenty different classification systems of these spectra had been proposed. But which was the right classification system? To understand the nature of this problem, it is helpful to consider the analogous problem of classifying people. Imagine a creature that has never before seen a human being—some sort of hypothetical alien, denizen of a distant world. Suppose it has been magically transported to Earth. It remains here for a mere few months, and during this brief interval of time, the creature minutely observes us as we go about our lives. And then it sets about the task of ordering people according to type.

Categories, of course, are utterly arbitrary and they are our invention. As the old joke goes, there are two types of people— those who insist on dividing people into types, and all the rest. Clearly there is an endless variety of possibilities our alien could adopt, and most of them would be meaningless. The creature might decide to classify people alphabetically. Would such a system be "right"? It has its uses (phone numbers and the like), but these uses do not correspond to properties of people in and of themselves. A good way to sharpen the point is to ask whether the adopted system is predictive—whether the alien could use it to determine in advance any *other* properties people might have. In this regard the alphabetical method of classification fails completely: people whose names begin with "A" are not all likely to be short, or intelligent, or wealthy. (A possible exception: a colleague with whom I discussed these matters ventured the opinion that people whose last names begin with "Z" would tend to be patient because they spend a lot of time waiting in lines. I'll leave this one up to the reader to decide.)

A more interesting system would classify people according to their strength. Type 1 people would be capable of lifting only the lightest of loads, type 2s could lift intermediate loads, and type 3s the very heaviest. Such a system has a certain amount of predictive power: people of type 3 would all tend to have beefy arms, for instance. But it too falls short, for with regard to other properties, it fails to cleanly separate people. As an example, it turns out there are two quite different sorts of type 1 people— those with very smooth skin, and those with very wrinkled skin. I'll explain why soon: till then, see if you can figure it out.

Ultimately the alien hits upon the following strange scheme of classification. The creature defines type A people as those who are weak, unable to thread a needle, and very short. Similarly type Bs are strong, able to thread needles, and taller; and Cs are weak, but able to thread needles and tall. Finally, the type D class is identical to the Type A except such people are tall.

Clearly a strange system. And yet the remarkable thing about it is that for all its artificial appearance, the ABCD classification reliably predicts a wide variety of human characteristics. Types B and C, for example, turn out on the average to earn more than Ds, who themselves have higher incomes than As. Similarly, there is a regular progression of the degree of smoothness of the skin, with As having the smoothest and Ds the most wrinkled.

Several lessons emerge from this analogy. In the first place, it makes clear that not all classification systems are equal: even though they are human inventions, some correspond more closely to objectively real properties of the objects under study. Second, neither the simplest nor the most obvious system need necessarily be the best—note in this regard the artificial and cumbersome nature of the alien's ultimate choice. Indeed, there is no way to ensure that we will ever come up with the best system. In the end it is a matter of trial and error.

And above all, it is a matter of a long-term immersion in the data. The slightest of clues may ultimately turn out to be the most significant. Annie Cannon ate, slept, and breathed stellar spectra. She knew each star's spectrum as an individual. Payne has given us a charming portrait of Cannon at work. The astronomical photographs with which Cannon worked were recorded on a glass plate. Such a plate would contain the images of perhaps hundreds of stars, each smeared out by a prism into a spectrum no more than an inch long, often faint and blurred. She would examine each spectrum through a magnifying glass, calling out her judgment to an assistant.

When I first looked at these plates [Payne writes in her autobiography], I was amazed. . . . It seemed impossible that anyone could see enough in those tiny smears to classify the spectra. Sometimes, indeed, I would find one of Miss Cannon's numbers in a spot where I could see nothing but a faint blur.

A legend was current that Miss Cannon could remember everything she had ever classified, and could immediately recall the serial number of the plate on which she had examined a particular star. But it is hard to believe that she could remember exactly *how* she had classified each of a quarter of a million spectra.

In the last years of Miss Cannon's life, [an astronomer] said "Somebody ought to find out from Miss Cannon exactly how she classifies each spectral type." I argued with him that she would not be able to tell them, because *she did not know.* She was like a person with a phenomenal memory for faces. She had amazing visual recall, but it was not based on reasoning. She did not think about the spectra as she classified them—she simply recognized them.

As a result of her deep familiarity with stellar spectra, Cannon was able to cut through the chaos of nineteenth-century astronomy, with its numerous systems of stellar classification, and discover the best. The system she created is still in use today. She then used this system in a project that was to become her life's work, the classification of every star in the sky bright enough to appear on Harvard's library of astronomical photographs. It was an extraordinary achievement: on a crystal-clear night, in which the stars scatter like jewels across the sky, perhaps two thousand might be visible to the naked eye—but she classified nearly four hundred thousand. What Linnaeus did for the world of organisms, Annie Jump Cannon did for the stars.

The secret of my hypothetical alien's ABCD system of classification is this: it describes a progression in time. People of type A are babies, those of type B are in the prime of life, Cs are entering old age, and Ds are the very aged. As for the alien's preliminary attempt at classifying people according to strength, it failed because both babies and the very old are weak.

Readers who have already figured this out will, I imagine, have been regarding my ABCD system with a jaundiced eye. It is indeed awkward, unnatural, and abstract. Worst of all, it conveys no understanding. Why waste time with such a thing when the natural and far more satisfying classification according to age is available? The reason is that this system may not have occurred to my hypothetical alien. After all, the creature was only granted a mere few months in which to conduct observations. Surely the correlation of the various categories with age is obvious to us only because we are in the process of living through them. Deprived of this insight, the alien might find itself stuck with the ABCD system, without the slightest comprehension of what the various categories meant.

In a similar fashion, we have no means of observing the aging of the stars. All the centuries since the scientific revolution amount to no more than a moment in cosmic terms. This was the situation facing astronomers when Cecilia Payne arrived at Harvard. Annie Cannon had codified the criteria to be used in classifying stars, creating a scheme analogous to the alien's ABCD system. But she had provided not the slightest insight into what the various categories might mean. In her Ph.D. thesis, Payne found that insight.

One might imagine, by analogy with my hypothetical alien's situation, that Cannon's spectral classifications actually corresponded to a progression in age, running from the youngest to the oldest stars. Many nineteenth-century astronomers, in fact, felt this would turn out to be the explanation. Indeed, from a few comments she makes in her autobiography, my guess is that Payne herself felt this to be the case at first. But once she got down to her research, she realized that this was not the case. A second possibility is suggested by the fact that Cannon's various categories of classification correspond to the signatures of various chemical elements. Her stars of type A showed evidence of hydrogen, while those of type K showed metals. Was it not natural to guess that A stars were composed of hydrogen, K stars of metals, and so forth? It was indeed natural—and it was, indeed, wrong.

The path that Payne followed to her insight into the true nature of Cannon's classification scheme was long and complex, and it led in the end to an unexpected conclusion.

The spectrum of the Sun cast upon the walls of Annie Cannon's house by the prisms of her chandelier ran the sequence of colors of the rainbow. So, too, with the spectra of stars cast upon photographic plates by an astronomical prism. In every case, however, each spectrum was interrupted by dark vertical lines. Cannon's classification scheme had dealt with these lines. They were called spectral lines, and they marked a set of colors absent from each spectrum. They were absent because, although emitted by the star, light of these particular colors was subsequently absorbed.

The light was being absorbed by atoms in the atmosphere of the star. Only certain colors were absorbed because an atom was capable of absorbing only certain energies, and the energy of light depended on its color. Payne knew that atoms could exist only in a restricted set of energy levels, much as a person standing on a ladder could occupy the lowest rung, the second rung, and so forth. The absorption of light by an atom would cause it to jump from a lower to a higher rung. But the light's energy had to be just right—too small would not be enough to move the atom to the next available level, while too great would attempt to move it one and a fraction levels, also impossible. Thus the spectral lines.

All this was relatively straightforward, and well known before Payne began her work. There was, however, a complicating factor: not all spectral lines were accessible to astronomical observation. The Earth's atmosphere was transparent only to a relatively limited range of colors—pretty much the range corresponding to visible light. But for many atoms, visible light corresponded not to a transition upward from the lowest level, but from a higher level. If these atoms did not happen to be occupying the right energy states, their spectral lines would lie in an inaccessible region of the spectrum. To Cannon, such lines would appear to be absent.

Cannon had classified stellar spectra by identifying the elements producing spectral lines. Payne showed, however, that the absence of a particular line was no proof of absence of the corresponding element. Rather, it simply meant that the atoms were situated at an inconvenient energy level. And what was it that determined whether an atom would be occupying the right levels—ones that produced the lines Cannon observed? The answer had been found by the Indian physicist Meg Nad Saha: it was heat. At low temperatures, all the atoms of a star's atmosphere would occupy their lowest energy states. But as the temperature was raised, they would progressively move up the ladder to higher and higher rungs. Only when the temperature of the star was just right would Cannon see the lines in its spectrum.

With this insight, the chain of explanation was complete. The Grand Fact that spectroscopists had discovered nearly sixty years earlier was now explained. It arose because some stars were hotter than others. As age was the principle underlying my hypothetical alien's system of human classification, so temperature underlay Cannon's. Her "O" category turned out to represent the hottest stars, with temperatures of 70,000°F. Her "M" category, the coolest, lay just over 4,000°F. For comparison, blast furnaces operate at about 3,000°F.

The essential insight had not been Payne's. It was Saha's. But Saha had not exploited it to the fullest. To implement his theory, it was necessary to know the actual pattern of energy levels for each chemical element. But when Saha did his work, this data was not available. By the time Payne began her thesis, atomic physics laboratories were beginning to provide the information—but Saha's isolation in India prevented him from obtaining their data.

Payne's Ph.D. thesis was a gigantic compendium, and it was a remarkable work. In it, she combined the very latest data with the most modern theoretical refinements to Saha's theory. On occasion she would turn the problem around, using astronomical observations to deduce what the energy level structure of some chemical element would have to be; experiments performed later in Earth-bound laboratories showed her inferences to be correct.

Reaction to her work was enthusiastic. One of the leading astronomers of the day, whose student had gone on to become director of the Harvard Observatory, wrote to say it was the best he had ever read (with the possible exception of the director's own). A review published years later termed it "undoubtedly the most brilliant Ph.D. thesis ever written in astronomy." And as for Sir Arthur Eddington, whose lecture back at Cambridge had inspired the youthful Cecilia to take her first steps in astronomy, he was a formal, reserved soul: he thought it "not so wild as might at first appear."

A group photograph has been preserved, taken the year Payne completed her thesis, of all the women at the Harvard Observatory (page 112). Fifteen of them stand before the ivy-covered walls and leaded glass windows. Annie Cannon gazes comfortably off into space. She is carefully dressed, with a brooch upon her chest and not one but two necklaces—she appears positively grandmotherly. Cecilia Payne is staring squarely at the camera. Her expression is serious, her dark hair combed neatly back.

Payne chain-smoked. She had little interest in clothing. She wore her hair short, was big-boned, unusually tall for a woman, and had a faint mustache. She was passionate about her work, and could be jealous of a colleague's success, a jealousy that was sometimes noticed. She thought of herself as shy and unattractive. Her daughter writes that she never saw her cry.

Payne could be formidable on occasion. But to those who knew her, this stern demeanor melted away. In the late 1920s a certain college student regularly dropped into her office for chats. Payne, at the time, was established in her field, widely respected, a mature scientist; the student, in contrast, was wet behind his ears. Nevertheless, she would talk with him for hours. She would talk about astronomy, about the arts, about the latest slight she had received from the observatory director. The student found her warm and sentimental, and with a wonderful sense of humor. She would quote from memory Gilbert and Sullivan patter songs. He would respond with a little T. S. Eliot, and the conversation would wander off into metaphysical reaches. This undergraduate was my father: I have rarely heard him speak of anyone with so much warmth and affection as he did in describing Cecilia Payne to me.

She was erudite in several languages, and addicted to puns in all of them. She would learn a new one for recreation. "Icelandic was a minor challenge," her daughter writes, adding somewhat unnervingly that "I cannot say she truly mastered it." Her conversation was rich in anecdote.

Payne was deeply affectionate, and would form passionate attachments to both men and women. These might become so intense as to resemble love affairs. She was capable of agonizing for weeks over a slight, real or imagined, and would suffer miseries if a recipient of her affections proved unworthy.

Many of her closest relationships were with other women at the observatory. No one has ever suggested there was anything remotely sexual about these attachments. But there does appear to have been a romantic tinge to her admiration for certain male astronomers. Often these would be authority figures—Eddington was one, and Harlow Shapley, the observatory director, another. Neither of these men was available: Shapley was married and Eddington a withdrawn, aloof individual.

What are we to make of these quasisexual attachments? It is easy enough to smile knowingly at them. Payne thought herself unattractive, and for many years seems to have felt that she would never marry. But to my mind this is too simplistic an approach. It demeans the importance of the intense admiration one feels for an acknowledged authority in one's field. These leaders are the ideals toward which every scientist strives. Such admiration very often spills over from the professional into the personal: in her case, the person in question was of the opposite sex.

Cecilia Payne's autobiography is often charming, but it suffers at points from a personal reserve, and from a syrupy sweetness that I find totally unconvincing. There is one area, however, in which all this drops away, and her recollections ring with bitterness and honesty: her descriptions of her experiences as a woman scientist. As a child she had to actively fight to get a scientific education—she refers to those years as "an uphill struggle." At one point, having declared an interest in science, her school principal openly told her she would be prostituting her gifts. On another occasion she won a school competition, the award for which was a book of her choice. Everyone thought she would opt for something like Shakespeare, but she asked instead for a book on fungi. Her teachers reacted with horror, doing everything in their power to dissuade the recalcitrant child.

At college she was regularly humiliated in class by the physicist Ernest Rutherford. Rutherford was one of the greatest scientists of his day, and by now he has attained the status of a legend. But his behavior toward Payne was contemptible:

> I was the only woman student who attended [his lectures] and the regulations required that women should sit by themselves in the front row. There had been a time when a chaperone was necessary but mercifully that day was past. At every lecture Rutherford would gaze at me pointedly, as I sat by myself under his very nose, and would begin in his stentorian voice: "*ladies* and gentlemen." All the boys regularly greeted this witticism with thunderous applause, stamping with their feet in the traditional manner, and at every lecture I wished I could sink into the earth.

Annie Cannon has left us no autobiography detailing her childhood experiences, but even though she went to a women's college, I am sure that she too faced much opposition in her love of science. What can account for these young women's determination in the face of such disgraceful behavior on the part of their teachers? What gave them the courage and determination to keep going throughout it all? On the one hand, it is obviously significant that both Cannon's and Payne's parents broke with tradition in having educated their daughters. And it may be significant that both grew up with strong and loving attachments to their mothers. Payne's father died when she was four, and aside from her brother, all the figures in her early life were strong women. And Cannon, as I have described, shared her early love of the sky with her mother.

In the early years, neither Payne nor Cannon was allowed to use a telescope to conduct regular observations. The feeble excuse was that it was dangerous for a woman to observe alone, and inappropriate for one to spend the night with a man. They were, by necessity, astronomers without a telescope. This practice was common at most observatories. Although Payne regularly taught courses at Harvard after obtaining her degree in 1925, not for twenty years did the university get around to listing them in the catalogue. Cannon did not receive a university appointment until she was in her seventies. As for Payne, even though the observatory director went to bat for her, Harvard's president explicitly told him that she would never have a position there so long as he lived. In fact, she did not receive one till her mid-fifties, when she was made professor—and at the same time, head of the astronomy department: this was the first time that Harvard ever admitted women as full members of the faculty. Both Cannon's and Payne's salaries lay far below those of their male colleagues.

"Young people, especially young women, often ask me for advice," Payne writes in her autobiography. "Here it is, *valeat quantum*. Do not undertake a scientific career in quest of fame or money. There are easier and better ways to reach them. Undertake it only if nothing else will satisfy you; for nothing else is probably what you will receive."

In later life, numerous honors did come to Cannon and Payne. But the question of honors for Annie Cannon is a difficult one. Her codification of our present stellar classification scheme was a major step whose importance I do not dispute. But this she achieved relatively early in her career. With minor exceptions, she did nothing more after that than classify stars— and I use the words "nothing more" here intentionally. In this task, Cannon was like a person who, every working day of her life, walks through a forest picking up leaves and identifying them—this one is an oak, that one a maple. The task required great diligence, but it was utterly repetitive and required scientific expertise in one narrowly defined field only. In her autobiography, Payne at one point mentions that a colleague of Cannon's "was always slowing things up by asking what it meant." It was a telling comment, for Annie Cannon never asked what it meant. It was Payne who asked, and answered, that question.

Cannon's life's work, the catalogue of nearly four hundred thousand stars, has proved an invaluable tool for generations of astronomers. But did this justify the bestowal of so many honors? Such a question is always a judgment call, and in answering it, I am vividly aware of the multitude of injustices she faced

in pursuing her career. Nevertheless, I must offer my opinion
for what it is worth: I believe it did not.

Cecilia Payne was one of the first woman astronomers to marry
without terminating her career. How this came about makes for
a curious tale. Beginning in the summer of 1932, three of her
dearest friends died in close succession. These multiple
tragedies forced her to the realization that she needed to im-
merse herself more fully in human affairs—by her own account,
she had to that point been so preoccupied with astronomy that
not even the stock market crash of 1929 had made much of an
impression on her. She resolved to live life more fully.

Promptly she fell in love, but unfortunately her love was not
reciprocated. Next year she embarked on a trip to Europe and
the Soviet Union. In Stalinist Russia, Payne was appalled by the
conditions under which the astronomers lived and worked. For
this reason, she was particularly receptive to the overtures of
the Russian émigré Sergei Gaposchkin, whom she met shortly
thereafter at a meeting in Germany.

"There are many wonderful anecdotes about Sergei that are
true," my father has commented, "and equally many that are
not." He was the sort of person about whom tall tales congre-
gate. The son of a Russian laborer, he was forced by economic
circumstances to leave school at seventeen. After numerous odd
jobs, he washed ashore—literally—in Turkey. Ultimately, he
obtained two Ph.D.'s, the first in literature and the second in astronomy. At the time he encountered Payne, Sergei's employment at a German observatory was threatened by the Nazis. He
pleaded with her to help him obtain a job in the United States.

Sergei thrust into Cecilia's hand a written account of his history. That night she read it. "I have not spent many sleepless
nights," she writes in her autobiography, "but that one was
sleepless. Perhaps this, I thought, is my one chance to do something for someone who needs and deserves it." Gaposchkin's
boss had already written Harlow Shapley, director of the Harvard Observatory, asking if a position was available. Shapley

had responded there was not. But Shapley had not reckoned on the determination of the woman who had written one of the best, not to mention the biggest, astronomical Ph.D. theses in history. Payne bombarded him with letters and, when she got back home, personal appeals. Ultimately Shapley surrendered, and offered Sergei a job.

Not long thereafter, Cecilia and Sergei married. Her colleagues were astounded. "I sincerely hope that it turns out splendidly," one colleague commented in a letter to a friend, "but I keep wondering how it happened."

Cecilia and Sergei collaborated scientifically for the rest of their lives. In this relationship, even though she was several years younger than Sergei, Cecilia Payne was always the senior partner. Her salary was at least twice his, and she was widely regarded by their colleagues as the better scientist. At home, much of the housework was left to Sergei. They had three children, whom they would bring to work. The kids roamed the observatory, exploring every nook and cranny. At one point the department formally voted to warn one of them not to disturb readers in the library. Cecilia was out of town at the time and reacted with high dudgeon: she took the vote as a personal insult.

In many ways Sergei was a hindrance to Cecilia's career. Historian of science Peggy Kidwell has written,

> [He] was rude to many who might have helped him, and undiplomatic with those who gave him opportunities. He was quick to take offense, yet spoke his own opinions with little regard for what others might think. He openly stated his admiration for female assistants at the Harvard observatory and expected them to be as impressed by his muscular physique as he himself was.

At one point when Cecilia was being considered for the presidency of a small college, his abrasive personality was mentioned as an impediment—ultimately, the position went to someone else. Many people feel that the scientific work that she did in collaboration with him was of inferior quality.

Sergei was a wild and wooly character who could be alternately exasperating and wonderful by turns. He wrote a three-volume autobiography entitled *Divine Scamper: Biography of Humanity.* The thing was published by a vanity press. It seems to have been cobbled together from a series of mimeographed, illustrated broadsides with which he would bombard friends. These were a marvelous hodgepodge of family matters, accounts of scientific discoveries, and personal reminiscences; some true, but others not quite so true.

In many ways Payne's choice of a husband made good practical sense. At the time, a woman who married was expected to abandon her career in order to further that of her husband. Numerous examples exist of female astronomers whose professional lives were thus terminated. But Cecilia's salary, low as it was, greatly exceeded Sergei's. They could not possibly have survived on his alone. So she was able to keep working. Furthermore, their relationship seems to have been in many ways a warm and loving one. Cecilia never allowed a word against Sergei in her presence, and she made numerous attempts to advance his career. As for Sergei, their role reversal must have been difficult for him, but it is greatly to his credit that he never showed it. "Cecilia," he once grandly announced to my parents, "is an even greater scientist than I am."

Shortly after completing her thesis, Cecilia Payne made a trip to England. There she met one day a famous astronomer. "Miss Payne?" he said dryly. "You are very brave."

In 1835 Auguste Comte had written that we would never learn the composition of the stars. Presumably they were made of *something,* he argued, but what this something was would forever remain unknown. It was an unfortunate prediction, for within a generation astronomers were routinely identifying in stars various chemical elements through their spectral lines. By the turn of the century, evidence of thirty different elements had been found in the Sun. Most notable of all was helium, which as its name suggests—*helios* is Greek for Sun—was first discovered in the Sun's spectrum, and only years later found on Earth.

By the time Payne began her work, it was possible not merely to identify the various elements in the distant stars, but to find out how much there was of each. After all, while the strength of a spectral line depended on whether the appropriate rung in the ladder of energy levels was populated, it also depended on how much of the element was present. Much of Payne's thesis was concerned with a detailed determination of the abundances of the elements in the stars.

She got into trouble over hydrogen. The simplest element, hydrogen is not particularly prevalent on Earth. But Payne found it to be extraordinarily prevalent in the stars. Indeed, she found them to be composed almost exclusively of this one element, with but traces of all the others.

But no one believed her. The astronomer she encountered in England had not believed her. Sir Arthur Eddington, the man who had launched her on her career and to whom she had once been attracted, did not believe her. Henry Norris Russell, mentor of her boss, did not believe her, and he told her so in a letter. "I am convinced," Russell wrote, "that there is something seriously wrong with the present theory. It is clearly impossible that hydrogen should be a million times more abundant than the metals." Faced with such opposition, Cecilia Payne backed down. She retracted her discovery, writing in her thesis that her enormous value for the abundance of hydrogen was "improbably high, and almost certainly not real."

There is nothing unusual about disagreements like these. It happens every day of the week in science, and it is the stuff of progress in the field. Arguments are the way science happens. But in this case, all of the power was on the opposing side. Eddington and Russell were among the leading astronomers of the world: Cecilia Payne was a graduate student.

And she was a woman.

In 1664 the first secretary of the Royal Society gave a definition of science, one that has always struck me as being quite astonishing. He defined science as "masculine philosophy." That was a comment of symbolic importance in social history, and it expressed an attitude toward female scientists that persisted for centuries. Maria Mitchell, the most famous American woman astronomer of the nineteenth century, worked part time as a librarian. Annie Russell, an astronomer at the Royal Observatory at Greenwich, resigned after marrying to assist her husband in his work. No one ever got around to paying her for her efforts. Caroline Herschel, who discovered eight comets, abandoned a career in astronomy to assist her brother William. She recorded his observations and fed him by hand when he ground glass for his telescopes. William went on to become one of history's most famous astronomers: Caroline's role has been almost forgotten.

Between 1859 and 1940, fully one out of every three American astronomers was a woman. I find this an utterly astonishing statistic, for women's names are conspicuous only by their absence from the textbooks and histories of astronomy. Why have these scientists been erased so thoroughly from the historical record? What made them so invisible? A large part of the answer is contained in a second striking statistic: half of these female astronomers had careers lasting less than five years. These women were astronomers only briefly. Whether they liked it or not, science for them was a temp job.

The emergence of this large class of invisible astronomers dates from the late nineteenth century, and it seems to have been caused by the invention of photography. Previously, astronomical data had been gathered by looking through the eyepiece of a telescope. It was slow work. But the new advance enabled a tremendous increase in the rate of gathering data. A single astronomical photograph would reveal literally hundreds of stars, all requiring careful study. The necessity arose of finding people to do this study.

There was no need for them to be fully trained astronomers, for their task was strictly limited. None had advanced degrees. These were the invisible women who swelled the ranks of astronomers, but remained forever on its periphery. Annie Cannon was one, and her work was characteristic of them all: limited, repetitive, and requiring no overall scientific competency. Indeed, such competency was actively discouraged by the astronomers—all male—who directed these technicians in their work. It would have decreased their productivity. "I always wanted to learn the calculus," one of these women commented, "but [the director] did not wish it."

These were the ladies of Observatory Hill, among whom Cannon and Payne stood for their photographic portrait. They were not the only female astronomers of the time: a few fully trained female scientists did exist, but there were never great numbers of them. As for the far more numerous technical assistants, historian of science Margaret Rossiter has compared them to assembly-line workers in a factory whose output was science. They were poorly paid; and just as the output of a factory brings few honors to its workers, so the scientific output of a department organized in this fashion brought little recognition to its technical staff. The system of classification Annie Cannon devised came to be known as the Harvard system, not the Cannon system. She stayed invisible. Nor was this system confined to Harvard, or even to the United States: it was widely practiced in Europe as well.

I have already mentioned that Cecilia Payne was one of the first women astronomers to wed. Indeed, many institutions maintained a rigid policy of *forbidding* their female teachers to marry. At one point a dispute erupted at Barnard College over whether a female physics instructor would be allowed to remain on the faculty after marrying. In spite of strong appeals by the department head, Barnard's dean refused. His rationale was that Barnard would not countenance presenting to its students the dismal example of a woman to whom home duties were of secondary importance. Male faculty, on the other hand, were actively encouraged to marry.

Use of the world's greatest telescope, the Palomar Observatory in southern California, was explicitly restricted to male astronomers until the mid 1960s. Prior to the Second World War, out of seventy-nine astronomers elected to the American Philosophical Society, only three were women. Not until 1978 was a female astronomer elected to the National Academy of Sciences. A defender of the status quo could always argue that the appropriate candidates were not available. I don't know whether this is true or not. But in any case, abundant evidence exists to show that female candidates for such honors were scrutinized far more carefully than their male counterparts.

On one occasion, a famous scientist explicitly inquired of a woman colleague, "If all the ladies should know so much about spectroscopes and cathode rays, who will attend to the buttons and the breakfasts?" This question, asked or merely implied, constituted the environment in which female scientists labored for centuries.

In such an environment, it is hard to imagine that Payne would have been believed in her assertion that the stars were made of little more than hydrogen. Furthermore, her claim ran counter to the prevailing temper of the day. Opinion and the herd instinct operate in science just as much as in other fields. In the long run, contentious issues in science are resolved in the court of objective reality: firm evidence will be found, irrefutable arguments developed. But scientists engaged in research do not live in the long run. They live now, and "now" is always uncertain.

The life of a practicing scientist is spent in a state of perpetual ambiguity. Partial evidence is uncovered, tentative arguments and conflicting insights. How one negotiates one's way through such a twilight state is a matter of personal taste—and of the subtle shifts in the winds of the day. At any given stage in the development of a body of knowledge, even though a firm conclusion has not yet been reached, prevailing opinion will usually be found favoring one particular view. So common is this phenomenon that scientists have a term for it: the bandwagon effect.

The researcher with a new idea is faced with an entire world that must be moved in order to win acceptance: a world of people convinced the idea is wrong; a world of evidence, hardly irrefutable but persuasive nevertheless, that must somehow be overcome. It is not an easy task—not for the most eminent scientist, and certainly not for a graduate student who also happens to be a woman. Furthermore, in spite of the common image of the heroic scholar battling in solitary splendor for acceptance of a revolutionary new idea, things do not happen this way very often.

It may not even be smart to push too strongly for your new idea. After all, it might be wrong! I'll go farther: *most new ideas are wrong*. In the task of exploring an unfamiliar universe, trial and error is the mode of operation. The person who invests years of time on one revolutionary notion is a person who has gambled on a long shot. Einstein did it and won. So did lots of other people. But should you?

What should you do if you think you have made an important discovery counter to prevailing opinion? I well recall a discussion of this question that developed over lunch a number of years ago. One colleague—a senior scientist well-versed in the politics and strategies of the profession—offered what struck me as the soundest opinion. He said that the best strategy would be to marshal all the evidence in your favor and make the case as forcefully as possible—and then you should quit. Get out of the field, he counseled, and leave the fight to others. In the long run, if your idea turns out to be false, you will not have wasted years of your life. And if it turns out to be true, people will remember who thought of it.

Sound advice, I have always felt—and it is just what Payne did. Although she seems never to have changed her mind, she made no further effort to persuade people that stars were made almost exclusively of hydrogen. Aside from everything else, her choice made another kind of sense, for certainly her thesis would never have been received so warmly, never have had so great an effect on the field, had she adopted any other course.

I believe that Payne could never have won this argument. All of the power was on the other side. But in saying this, I do not wish to imply that Eddington and Russell and all the other astronomers who opposed her were wrong in having done so. In fact, they had good reasons for their positions. At the time Payne's thesis was published, Eddington was putting the finishing touches on his theory of stellar structure. The idea that stars were composed primarily of hydrogen did not mesh with this theory. Eddington had developed a whole list of arguments, all pointing to the opposite conclusion of a low hydrogen abundance. None of these arguments was convincing when taken alone, but they were impressive in their number and variety. Furthermore, such a low abundance, far from constituting a peripheral element of his theory, was intimately woven into its fabric in a number of ways.

And Eddington wielded enormous influence. He was aristocratic, reserved, handsome, elegant—he was eventually knighted.

And above all, he was brilliant. Few of his contemporaries were capable of creating so comprehensive an edifice as his theory of stars. Therefore, they were unwilling to challenge it. People had gotten into the habit of simply listening to what he had to say.

As for Russell, he was perfectly well aware that Payne had some arguments on her side. But he had more on his. As early as 1914 he had published a paper arguing that the abundances of the elements showed a striking uniformity across the entire astronomical universe. The chemical composition of the Earth looked to be pretty much the same as that of the Sun. And the Sun's, in turn, was similar to that of the meteorites that fell from the outer reaches of the solar system. The notion of a universal composition was in the air: the heavens were made of the same stuff as the Earth.

And Payne was claiming that the heavens differed from the Earth—differed fundamentally, differed in their very substance. But her arguments were far from ironclad. A fully detailed analysis of the abundances of the elements would require the full theory of atomic structure, which at the time was still in its infancy. She had taken shortcuts, made assumptions, that only further work could test.

In the end, Eddington and Russell turned out to be wrong, and Payne right—and, remarkably enough, it was Russell who showed this. Over the next four years, he worked to develop the full mathematical theory of spectral lines, and eventually he found himself forced to side with Payne. In 1929 he published a magisterial paper on the composition of the Sun. In it he showed that she had been correct in her claim, and he referred to "the very gratifying agreement" between his new conclusion and hers. He gave in his paper full credit to her prior work—but as historian Peggy Kidwell has emphasized, nowhere did he so much as mention his role in having once persuaded her to back down.

In the course of working on this profile, I found myself faced with a curious situation. I was finding it difficult to pay attention. Instead, my mind was continually wandering. Eventually, I realized that it kept wandering to the same topic—to the photograph of the ladies of Observatory Hill reproduced in Figure 3. Why was this picture so regularly recurring to my thoughts? Ultimately, I realized that a particular element in that apparently prosaic image had been insistently nagging for my attention: the presence of Cecilia Payne.

The photo, indeed, presents us with yet another instance of classification; in this case, one according to sex. But was that an appropriate classification? Everyone seems to have thought so.

But while Annie Cannon fits comfortably into this portrait, Cecilia Payne does not. In an important sense she was not a member of the group in which she stood. Properly speaking, she should have taken her place in another group photograph, that of the professional astronomers—the men. But no one saw fit to place her there.

Cecilia Payne was a transitional figure in the history of women in science. Annie Cannon was of the old breed, and she was comfortable with her place. Cecilia Payne was the first of the new. Decades later, on the occasion of her being appointed head of the astronomy department, she invited all the women students to a celebratory tea. "I find myself in the unlikely position of a thin wedge," commented this massive, sturdy woman, to gales of laughter from the happy crowd.

2

The Bulldog

Ludwig Boltzmann and the Second Law of Thermodynamics

Upon accepting a university appointment at Graz, it occurred to Ludwig Boltzmann that he'd like to live the country life. So he bought a farmhouse several miles out of town. Deciding one day that no farmhouse would be complete without a cow, he went down to the local cattle market to buy one. Torn shoes flapping, stout, perhaps a bit otherworldly, the great physicist led the beast home through the town. For information on how to milk it, he had already consulted a professor of zoology.

A charming image. Boltzmann was, in fact, a charming man—earnest, modest, and open in his personal affairs. He was a man of many passions, and his soul was continually swelling with one emotion or another: with tears at beauty, with pride at accomplishment, with anger at theories he regarded as misguided. In his professional life, on the other hand, Boltzmann presented a different aspect: formidable, severe, a master of the most difficult and complex of mathematical calculations, and an indomitable battler.

He battled for acceptance of the atomic hypothesis, the notion that everything is made of atoms. While we accept this without thinking too much about it nowadays, the notion was in plenty of doubt during Boltzmann's lifetime—there were, in fact, a number of reasons to suspect it. But Boltzmann's triumph was to show that the hypothesis allowed him to explain things in beautifully simple terms. Again and again throughout his career, he returned to this task. Watching him in debate, a colleague once was reminded of a bulldog: tenacious, unrelenting. It is an image that aptly sums up his entire scientific style.

Boltzmann concentrated on thermodynamics—the study of the flow of heat. Its content is summarized in two laws. The first law, developed in fits and starts throughout the first half of the nineteenth century, states that heat is a form of energy. The notion can be traced back to Count Rumford, the inventor of the Rumford fireplace. Rumford was an American who fled to England to avoid the Revolutionary War; he later became a military engineer and was employed by the Bavarian government. Impressed by the great amount of heat generated in boring cannon, he instituted a series of measurements and decided that heat must be some form of motion. The climax of the development of the first law was a paper read by the German physicist Helmholtz before the Berlin Physical Society, in which he announced what we today regard as one of the cornerstones of science: the law of conservation of energy. The paper was rejected by the scientific journal to which he submitted it, and Helmholtz ended up publishing it privately as a pamphlet.

But if heat is a form of energy, a paradox arises—for energy is hard to come by, but heat is everywhere. Even on the coldest of days, we are surrounded by at least some warmth. The heat energy of air, for instance, is quite impressive: the quantity contained in a few cubic yards of it would be sufficient to accelerate an automobile to 60 miles per hour. In the process of accomplishing this feat, those few yards of air would be cooled down to absolute zero—but once the car arrived at its destination and braked to a halt, its energy of motion would be converted back into heat: after all, when brakes are used to slow a car, its energy of motion warms the brake pads, which ultimately warm the air. It turns out that the air would be warmed right back to its original temperature. So the journey would have been accomplished with an expenditure of no energy whatsoever.

Nor would anything prevent us from repeating the process indefinitely, continually draining energy from the atmosphere and then returning it, and so journeying on forever. The first law of thermodynamics implies that we should be able to build such a perpetual motion machine. A machine like this would be an instant solution to the world's energy problems. Never mind all those costly hydroelectric power plants, the coal plants with their coal dust and the nuclear reactors with their radioactivity. Free energy floats around everywhere, as common as air—it is *in* the air. All we need do is scoop it up.

Unfortunately, such a perpetual motion machine has never been built. People—many people—have tried and failed. But why have they failed? If the first law allows it, some other principle must intervene to forbid it. What could that principle be? Here is a clue to its nature: prior to the transfer of energy from air to automobile, the two must have been at pretty much equal temperatures—but during the transfer, they would have developed *unequal* temperatures. The air would have gotten colder and colder, while the engine grew hotter and hotter.

But this is the sort of flow of heat that never happens. In reality the flow is always in the other direction. Things initially at unequal temperatures tend to reach the same temperature. If an ice cube is dropped into a glass of warm water, the ice melts while the water cools off. But the heat flow envisaged in the perpetual motion automobile would be just the opposite. It would be like dropping an ice cube into a glass of water and finding the water growing yet hotter while the ice cube grew yet colder.

Only if heat flowed in the wrong direction could perpetual motion exist. The principle that forbids this is the second law of thermodynamics. This law was formulated in modern form by Rudolf Clausius in 1865. Clausius mathematically defined a certain quantity and then showed that the impossibility of a reverse heat flow followed from the principle that this quantity could only increase. As for what his new quantity was to be called, Clausius was of two minds:

We might call it the *transformational content* of the body. . . . But as I hold it better to borrow terms for important magnitudes from the ancient languages, so that they may be adopted unchanged in all modern languages, I propose to call [it] the *entropy* of the body, from the Greek word "trope" for "transformation." I have intentionally formed the word "entropy" to be as similar as possible to the word "energy"; for the two magnitudes to be denoted by these words are so nearly allied in their physical meanings, that a certain similarity in designation appears to be desirable.

The second law of thermodynamics is the principle that entropy can never decrease. It is what makes heat flow in the right direction—and it is what keeps us from building a perpetual motion machine.

Boltzmann loved the atomic theory of matter. He loved it because it explained things, simply and easily. According to it, heat was a form of motion; the hotter the body, the more rapid the motion. If the moving thing was large—a galloping horse—people were accustomed to describing the motion in terms of energy. But if the moving thing was too small to be seen—an atom within the horse—no one noticed the fact that it was moving, and people contented themselves with speaking in terms of heat. Notice that, according to this view, the first law of thermodynamics has become obvious, indeed trivially so. It has completely lost its status as a separate law of nature.

Boltzmann aimed to do the same with regard to the second law. His progress toward this triumph has been traced in a fascinating article by historian of science M. J. Klein. Boltzmann made a first cut at the problem the year after Clausius announced his new law of nature. His paper begins by discussing the status of the first law, which "for a long time already" had been known to follow from the atomic theory. The "long time" amounted to less than twenty years, but Boltzmann himself was only twenty-two at the time—and indeed, one senses throughout his paper the impatient voice of youth. "It is the purpose of this article to give a purely analytical, completely general proof of the second law," he grandly announced, apparently failing to notice that he had done no such thing.

He did, however, succeed in deriving a result of more limited importance. Five years later Clausius himself published a paper in which, making no reference whatever to Boltzmann's work, the same result was obtained. During the intervening years, Clausius had found himself pressed for time, and he seems not to have bothered to keep up with what was going on in the field. Boltzmann's response, though, was hardly a model of tact: he published a paper pointing out his priority, reproduced much of his original article as proof—publishing standards seem to have been lower in those days—and imperiously concluded "I think I have established my priority. I can, finally, only express my pleasure that an authority with Mr. Clausius's reputation is helping to spread the knowledge of my work."

In 1871 he returned to the subject, and managed to find a formula for the entropy of a gas expressed not in Clausius's terms, but in new ones—in terms of the state of motion of its constituent atoms. And in the following year he managed to prove that the entropy, as defined through his new formula, could never decrease.

He had proved the second law. It had been six years since his first try. Boltzmann seems to have been almost diffident about announcing this remarkable triumph—it is tucked away in the middle of a paragraph, thirty pages into a hundred-page paper forbiddingly entitled "Further Studies on the Thermal Equilibrium of Gas Molecules."

Ludwig Boltzmann's skull was large and strong. He had what people of his day might have called a philosopher's brow. His face was flushed, with a large nose and thick ears. His beard was thick, and he had masses of curly brown hair. Boltzmann was very nearsighted, and all his adult life he peered out at the world through tiny, Trotskyesque glasses. A delightful drawing of him by one of his students (see page 113) shows a stout figure, not particularly tall, and slightly stooped—sort of a Father Christmas figure. He is gesturing earnestly as he explains a difficult point of mathematics. In another drawing Boltzmann bicycles down the street, wobbling erratically, coattails flying.

He had been born in 1844 in Vienna, the eldest son of a revenue official. His early years were marked by tragedy: both his sister and his younger brother died at an early age, and his father died when Boltzmann was fifteen. He was educated at home by a series of tutors, one of whom was the composer Anton Bruckner. Bruckner had not lasted very long: arriving one day in the midst of a rainstorm, he had tossed his wet coat upon a bed. Frau Boltzmann remonstrated, and the composer stormed out in a huff, so ending this phase of young Boltzmann's musical education. In later life Boltzmann grew quite accomplished on the piano, and would often play at musicales—"with rousing fire and with secure technique," according to one observer.

Boltzmann was a wizard of a mathematician and a physicist of international renown. The magnitude of his output of scientific work was positively unnerving. He would publish two, three, sometimes four technical books a year; each was forbiddingly dense, festooned with mathematics, and as much as a hundred pages long. He was what we would now call a hot property, and was continually receiving job offers. He bounced from university to university throughout Germany and Austria.

"The stars bend like slaves," he wrote, "to laws not decreed for them by human intelligence, but gleaned from them." The gleaning of these laws, and the development of detailed, mathematical theories, was his greatest love. He lived in an age in which theoretical physics was just being recognized as a separate discipline. A document exists from this period in which the faculty of the University at Munich find it necessary to describe to their administration the growing specialization of physics, and the need to recognize theory as a field in itself, in order to justify their proposal to hire Boltzmann specifically as a theoretician.

Boltzmann did, in fact, spend many years at Munich. For a time there he met weekly at one of its famous beer halls for chats with a group of scientists and engineers. Although his research interests were highly abstract, he also had a strong practical bent, and he was fascinated by technology. For a time he gave financial support to an inventor who, long in advance of the Wright brothers, was working to develop heavier-than-air flight. (Boltzmann gave a number of lectures on the subject, complete with model airplane.) For electricity he foresaw a great future:

[P]erhaps the time is not far off when every household bill will celebrate those great electricians Ohm and Ampere; and perhaps in the coming century every cook will know how many watts are needed to roast meat, and how many ohms her lamp has.

When a colleague invented a new lightbulb, Boltzmann invited members of the Vienna Physical Society to his home for a demonstration. Sandwiches and 50 liters of beer were set out: seven people showed up. He built by hand, and entirely from scratch, an electric sewing machine for his wife.

In spite of his many successes—he was one of the most prominent physicists of his time—he was modest and unassuming in personal conversation, and a man of great charm. Students and colleagues alike were immediately put at ease upon visiting his home. He was widely known to be a soft touch for a favor, and would pace the streets miserably for hours if forced to flunk a student; toward the end of his life, he never could bring himself to flunk anybody. He was a ready weeper, and would cry openly at the railroad station whenever circumstances forced him to leave his wife for long.

Boltzmann was subject to recurring depressions, and sometimes in company would lapse into long dismal silences from which he could not be roused. He himself would comment upon these intense swings of emotion from great joy to deep grief. In the light of modern knowledge, it seems that he suffered from manic depression: he committed suicide in 1906.

By 1872 Boltzmann felt he had proven Clausius' second law of thermodynamics. He believed he had demonstrated that, like the first law, it followed from the hypothesis that things are made of atoms. Unfortunately, his proof could not possibly have been correct, for if you believe that atoms exist, then you will have to conclude that the second law is false. It was the British physicist James Clerk Maxwell who had realized this, and he appears to have first mentioned his reasoning in a letter to a friend in December 1867. In this letter Maxwell outlined a specific method by which heat could be made to flow the wrong direction, thus reducing the entropy and so violating the law.

Maxwell's argument shows how to build a perpetual motion automobile. It is a sort of Stanley Steamer, though with improvements. Its construction is remarkably simple: the engine consists of nothing more than some air in a bottle, one wall of which is pierced by a tiny hole. The hole provides a contact with the surrounding atmosphere, although it is so small that only one atom at a time can pass through. Maxwell showed how, as a result of this passage, the air in the bottle could be made to grow steadily hotter and hotter—ultimately hot enough to boil water and so propel the car.

The atomic theory visualized both the atmosphere and the air in the bottle as composed of atoms. They were always moving. As a result, even though there was no net flow of air through the hole—no wind— there was an incessant, invisible interchange. Every so often an atom from inside the bottle happened to find its way out through the hole: every so often an atom of the surrounding air passed in.

A further element of the atomic theory was that the hotter the gas, the more rapid the motion. So the atoms within the bottle moved faster than those on the outside. Whenever one passed outward through the hole, it therefore carried energy outward and so cooled the engine; whenever a slow one from the outside air passed in through the hole, it cooled the engine still more. In this way the atomic hypothesis explained why hot things cooled while cold things warmed.

The flaw that Maxwell saw in this explanation revolved around a discovery he had recently made: that the atoms making up a gas did not all move with the same velocity. Even a cold gas contained the occasional rapidly moving particle; even a hot gas, a few nearly at rest. In his letter Maxwell suggested a means whereby this fact could be used to reverse the flow of heat and so violate the second law. He proposed putting a door on the hole connecting engine to air, and a gatekeeper beside the door. The job of this gatekeeper would be to watch the atoms heading toward the hole and to decide which to let through. Simply by opening and shutting the door, it could arrange that the only particles that filtered outward from the engine were those few slowly moving ones. Similarly, it could ensure that only the fast ones filtered inward. In this way, the gatekeeper could arrange for heat to flow the wrong way, heating the already hot engine more and more.

In a subsequent letter Maxwell described his gatekeeper as "a doorkeeper, very intelligent and exceedingly quick, with microscopic eyes." Nowadays we call it Maxwell's demon. Simply by the application of knowledge, such a demon would be capable of violating the second law of thermodynamics. The prettiest thing about Maxwell's creation is the reinterpretation it introduces of entropy. Clausius's definition of this quantity had been highly mathematical and abstract, and had provided little insight into its underlying nature. Maxwell's achievement was to show that entropy is related to information: the information that the demon possesses about the velocities of the atoms of the gas.

More precisely, entropy is related to information's absence—to confusion. In most circumstances the physical universe is "stupid": its operations take place without the benefit of an ordering intelligence. But the function of Maxwell's demon is to impose order on this random and generally messy world, and it does this by being smart. The low-entropy state it acts to bring about is a highly structured one in which all the atoms of a gas move with more or less the same velocity. Conversely, the increase of entropy corresponds to the lapse into disorder that follows once the demon's hand is lifted. The second law of thermodynamics describes the general decline into messiness of the universe. It says that the children's room is guaranteed to become messier if they don't keep picking things up.

Boltzmann did not immediately respond to the conundrum posed by Maxwell's demon. Indeed, Klein emphasizes that Boltzmann appears not even to have known about it at the time he formulated his "proof" of the second law. The lapse is all the more remarkable in that Maxwell's argument was spelled out in his popular textbook of thermodynamics, a book that had been available for five years by 1872. It was not Maxwell but Josef Loschmidt, a colleague and friend of Boltzmann, who ultimately forced him to return to the subject.

Loschmidt had been born in poverty and had tended cattle in his youth; in middle age, while earning a living as a grade school teacher, he made a famous calculation for which he was granted an honorary degree and a faculty position at the University of Vienna. Loschmidt could hardly be accused of not believing in atoms—he was the first person to determine their sizes. Nevertheless, he raised an objection to Boltzmann's work, and it was one that struck at the very heart of the program of the atomic theory of matter.

I have always found Loschmidt's objection stunning, both in its simplicity and in its force. In a mere few steps, he seemed to prove that things could not possibly be made of atoms. His argument made no mention of the mathematical details of Boltzmann's "proof." Rather, it related to something that at first sight seems unrelated: the one-way nature of the flow of time. Time always flows forward, never backward.

What does it mean to say that time flows backward—or forward, for that matter? The notion can be made precise by the following imaginary project. Imagine making a film of some sequence of events. As for the subject of the film, it could be anything. It might be water flowing in a brook; it might be a car crash; it might be a swinging pendulum. But let us now show the film to a friend, and let us arrange that the film can be threaded through the projector either forward *or backward*. The films that we chose to thread in the normal fashion would show the normal flow of time—but those we chose to thread backward would illustrate what it would look like were time to flow backward.

Our friend's task is to guess which choice we made, forward or backward. Usually, it will be easy to make that guess. After all, most reversed films would look nonsensical. One would show water flowing uphill. Another would show a tangle of metal and glass scattered about on a pavement, a tangle that suddenly and magically rises upward to assemble itself into two intact automobiles, which then speed backward away from each other into the distance. Such things, of course, never happen.

And yet a third such time-reversed film would show heat flowing in the wrong direction. It would show entropy decreasing. That never happens either.

This imaginary project dramatizes the irreversible nature of the passage of time. But remarkably enough, it only occurs in certain circumstances. A whole class of phenomena don't show such irreversibility at all—the phenomena treated by the science of mechanics. A film made of a swinging pendulum, for instance, would show nothing more than the same swinging pendulum if shown in reverse. So, too, with one of a planet orbiting the sun, or of the collision of two billiard balls. All these purely mechanical processes can occur "backward" as well as "forward"—and indeed, the distinction between the two terms has little meaning in these examples. Our friend, if shown a film of any of these, would have utterly no way of knowing whether we were playing it forward or backward.

Entropy is time's arrow. It is entropy that distinguishes processes that pick out a direction to the flow of time from those that do not. The flowing of water, the collision of two automobiles, and the normal flow of heat are processes in which the entropy increases. But there is no such thing as entropy for a swinging pendulum, nor for a planet orbiting the sun or for the collision of two billiard balls.

Nor is there any such thing as entropy for the collision of a lot of billiard balls. But a lot of billiard balls is, if you believe in the atomic hypothesis, all there is to matter. Loschmidt's objection to the theory that matter is made of atoms is that according to this theory, there would be no difference between time flowing forward and time flowing backward.

Loschmidt was arguing that nothing in the physics of the motion of atoms picks out a direction to the flow of time. The motion of these atoms is analogous to the motion of billiard balls. The popping of a balloon, for instance, corresponds to the sudden rushing outward of a large number of atoms originally confined to a small region of space. It is analogous to the opening of a billiards game, in which the initially neat arrangement of balls placed in the rack is shattered. The reversal of this motion is actually possible, and it would cause them to reassemble into precisely their original pattern: according to the atomic hypothesis, balloons could unpop just as easily as they could pop.

For every atomic motion that increased the entropy, Loschmidt could point to another that decreased it: the reversed motion. Thus, he argued, there would be no general tendency for the entropy to increase, and Boltzmann's program of comprehending the second law of thermodynamics was doomed to failure. And more than that—the concept that matter was made of atoms could not be reconciled with the irreversible nature of the flow of time.

Maxwell's and Loschmidt's objections both describe ways in which the second law of thermodynamics can be violated. It is worth noting that both objections involve knowledge and intelligence—with taking care. Maxwell's demon would have to scrutinize each approaching atom in order to decide which to let through the gate. And to reverse the motion of each escaping atom from a popped balloon, a similarly encyclopedic knowledge would be required.

But Boltzmann, finally forced by Loschmidt's argument to return to the problem, focused on yet another means of decreasing entropy: not through the application of knowledge, but simply through waiting. I have already emphasized that the second law of thermodynamics describes the general decline into disorder of the world: that the children's room is guaranteed to become messier if they don't keep picking things up. Boltzmann, however, added a new element to this analogy. He noticed that if children pick up and put down their clothing randomly, once in a very long time they will manage to toss everything into the closet just by the law of averages. If you wait long enough, order can appear. And similarly, if you wait long enough, a situation will arise in which a balloon will unpop. Not only that, but once in an eon the few rapidly moving air particles in the chilly atmosphere will make their way into the engine of a perpetual motion machine and so heat it.

Both of these processes involve a change of entropy in the wrong direction. Boltzmann's conclusion was that this is quite possible. So the second law of thermodynamics must possess a character quite different from that of every other natural law. Previously, the laws of physics had been taken to be absolute: unsupported bodies invariably fell, and energy was always conserved. But entropy only increased most of the time.

Boltzmann's reaction to this recognition was a remarkable creative shift. He developed an entirely new way of thinking about the second law, and it is one that has always impressed me as being of marvelous beauty and great subtlety. He thought of it not in terms of collisions among atoms, but in terms of randomness and probability.

Boltzmann analyzed how *likely* it would be for the entropy to decrease. He showed that the state of greatest entropy was the most probable state; and that while other distributions corresponding to a smaller entropy were possible, they were overwhelmingly less likely. Because of the very great number of atoms involved in any real gas, sequences of decreasing entropy would never be observed in practice. Thus our daily experience of the irreversible nature of the passage of time was not in contradiction with the atomic hypothesis. While it was indeed true that heat could briefly flow in the wrong direction and that balloons could unpop, such sequences of events were so rare that they had almost certainly never happened in the entire history of the Earth.

But everything not forbidden is compulsory. Given enough time and enough space, even the most unlikely event is guaranteed to happen. Loschmidt's insight into the nature of time led Boltzmann to speculations about the universe as a whole.

Somewhere in the cosmos, he argued, there must exist isolated pockets in which events were taking place in the wrong direction—in which time was running backward. Entire galaxies could be involved in this process. What would it be like to watch such galaxies through a telescope? Boltzmann's first thought was that we would see water flowing uphill, shattered automobiles reassembling themselves, and the like. We would see all the crazy things illustrated by a time-reversed movie. But he quickly realized that we would never be able to see such galaxies. They would be invisible—because their stars, rather than emitting light, would be steadily absorbing it, the light pouring inward upon them from the rim of the universe in the time-reversal of the normal flow. Such galaxies would be entirely hidden from our view. And if we could send a message to the inhabitants of such a backward-running world, they would forget it the instant after they received it, for their memories would be running backward in time. Indeed, they would "remember" the message up to the instant it arrived.

But as the entropy decreased in such isolated pockets, it nevertheless continued increasing in the universe as a whole. The cosmos, then, must have begun in a state of unusually low entropy, and has been evolving toward more disordered states ever since. Such speculations led Boltzmann to a new view of the origin of the universe. He came to think of this origin, not as an act in which matter was created, but as an act in which order was imposed upon already-existing matter. The clock of time was set to zero, and the universe set forth upon its downward-running course, by a cosmic Maxwell's demon—a Maxwell's god. Creation was an act of intelligence.

Seen from the standpoint of his new conception of the meaning of entropy and the law of its increase, Boltzmann's 1872 "proof" of this law must have been wrong. But in his response to Loschmidt's argument, not once did he acknowledge this. Furthermore, Boltzmann may never have come to his reinterpretation of the second law of thermodynamics, and the radical extension of its scope, had he not been forced to do so by Loschmidt. But he never acknowledged his debt.

Indeed, Klein emphasizes that everything about this episode is marked by miscommunication and confusion—by high entropy. As Clausius had not known of Boltzmann's work, and Boltzmann had not known of Maxwell's, so most physicists of his day failed to come to grips with Boltzmann's more sophisticated analysis of the second law. As late as twenty years after he had effectively answered Loschmidt's objection, a new attack on Boltzmann's work was launched by Ernst Zermelo, who formulated a paradox closely akin to Loschmidt's. Page after page of the scientific journals were filled with discussion of Zermelo's paradox, in spite of the fact that it can be resolved in precisely the same way that Boltzmann had resolved Loschmidt's.

How can we understand this persistent miscomprehension on the part of the physics community? It cannot be the result of simple lack of recognition, for Boltzmann was widely regarded to be one of the foremost physicists of his day. Part of the reason must lie with the nature of his analysis, which I can testify from personal experience is of unusual subtlety—you think you have it, but then it slips away from you. Part must also lie with Boltzmann himself. His papers were wordy and convoluted, and he never seemed able to say a thing in one word if ten would do. Coupled with his immense scientific output, his writing style led a number of his contemporaries to throw up their hands in despair. "By the study of Boltzmann I have been unable to understand him," complained Maxwell in a letter to a friend. "He could not understand me on account of my shortness, and his length was and is an equal stumbling-block to me." Maxwell appears to have totally stopped reading Boltzmann's papers by about 1870.

In the summer of 1905 Boltzmann took a trip to America to give a series of lectures at Berkeley. "Voyage of a German Professor to Eldorado" is his account of that trip: it makes for wonderful reading.

The account suffers from the endless wordiness that so maddened Maxwell. Boltzmann tells us in detail what he had for dinner on the night of his departure, what the band played during the ocean crossing, and far more than we would care to know of a growth that developed in his armpit. On the other hand, Boltzmann's style is courtly and exquisite, and it is often poignantly beautiful.

And at times the most delightful humor breaks out. He pays a visit to the Stanford campus, built in memory of the Stanfords' son who had recently died. "In Europe," Boltzmann comments, "when such a tragedy happens, a woman goes out and buys herself a dozen cats or a parrot. These people built a university." During a visit to a laboratory of artificial insemination, Boltzmann muses on what a difference to society the technique would make if applied to humans. Women could drop down to the corner store to buy little bottles of liquid—one for boys, another for girls. They would no longer need men, "and at this point, wine would become totally obsolete."

Boltzmann, in fact, loved wine, and at one point in his trip he got into trouble over it. Like any other traveler to depraved and exotic lands, he developed diarrhea—from the drinking water in Berkeley. Apparently, it was mostly rainwater, kept in huge cisterns, and it made him deathly ill. He resolved to drink only wine, and asked a fellow faculty-member to guide him to the nearest liquor store.

He could not have made a more appalling request. The reaction was utter shock. Boltzmann comments that he might as well have asked directions to a house of prostitution, and his colleagues at Berkeley shrank from him in consternation. He eventually located a store, but he felt it more politic to drink alone in his room. "This they call freedom," he grumbled sourly in a letter to a friend.

Everything about Berkeley fascinated Boltzmann—the number of women about (his fiancée had vainly struggled for years to enroll as a university student in Austria); its governance (Mrs. William Randolph Hearst, the newspaper baron's widow, controlled it absolutely, passing on to its president her decisions); and its heating system (hot water piped through metallic, box-like structures; radiators were apparently a rarity in 1905).

There is also much condescension in Boltzmann's account of California. In many ways his attitude is that of an aristocrat strolling through the hustings: everything is interesting, but nothing quite measures up. During his trip across the country, he is thrilled by the sight of the Sierra Nevada through the train window, but he takes care to remind the reader it is no Semmering. He is invited for the weekend to Mrs. Hearst's estate, surrounded by "a fantastic gate, and not that ugly." There at an after-dinner musicale a music professor performs on the piano: "he played well," Boltzmann notes, "and even knew that

Beethoven wrote nine symphonies, and that the ninth was the last."

On the other hand, Boltzmann often found himself impressed by California, both by its natural beauties and by the spirit of the place. If Mahler could only compose a symphony evoking the glories of Yosemite "the trees there would have shivered for joy"—although he cannot resist the dig that the people there would not have understood it. Standing at the base of the Lick Observatory, named in honor of the wealthy philanthropist who had financed it and even then lay buried beneath it, he muses of "two idealists, Schiller and Lick. Schiller would be saying 'I don't need you.' But Lick would prove him to the contrary." "Happy the nation in which millionaires think idealistically, and idealists become millionaires."

Throughout Boltzmann's account, the reader encounters a man passionately engaged in his travels. Here is someone in love with life and who takes it to the fullest; who appreciates good food and wine and the company of women; and who can write of his ocean voyage

> I wept when I saw the color of the sea—how can a mere color make one cry? Or moonlight, or the luminescence of the sea in a pitch black night? . . . But if there is one thing which is more worthy of our admiration than natural beauty, it is the art of men who have conquered this never-ending sea so fully in a struggle that has been going since the time of the Phoenicians.

And yet this same man was to die by his own hand within a year.

Boltzmann loved the atomic hypothesis because it explained things. It proposed a striking and easily visualizable mechanism underlying thermodynamics, and while the mathematics of the theory may have been complex, the essential idea was beautifully simple. He argued that only through such mechanistic accounts could true understanding be reached. They allowed us to visualize the means whereby things worked—the actual nuts and bolts of the universe: what happened inside. If you knew the mechanism you knew everything. It was the very stuff of explanation.

But they are not the stuff of explanation in physics anymore. One of the great shifts in thought of physics has been the recognition of the essential insufficiency of mechanistic theories of nature. This shift took place toward the end of the nineteenth century: it is the tragedy of Boltzmann's life that he had the bad luck to be caught in the tide.

I can still recall as a student reading of a nineteenth-century theory that tried to explain the mechanism underlying the forces between electric charges. The author of this theory asked us to imagine two balloons immersed in water, both expanding and contracting in a regular oscillation. Because of this oscillation, each balloon produced a series of waves through the water, which then exerted forces upon the other. It turned out that if the two balloons oscillated synchronously, the net result would be a repulsion between them. But if one expanded while the other contracted, the force would be one of attraction.

I was delighted by this ingenious explanation of the attraction of unlike charges and the repulsion of like charges. But by now this theory has fallen into oblivion. No one remembers it any more. The problem is that it fails to explain anything *else* about charges—for instance, why they develop magnetic properties if set in motion, or how they emit electromagnetic waves. In Boltzmann's time the most elaborate models of electromagnetism were proposed to fill in these gaps. In one, space was envisaged as being filled with ether, an invisible substance imbued with the most marvelous properties; in another, with a network of rods, levers, gears, and countergears. Each model was more ingenious than the last: none of them has stood the test of time.

With each failure, the implication grew stronger that the phenomena of electromagnetism could never be explained in such terms. This was simply not how electromagnetism worked. The very program of searching for mechanism in nature came to be regarded as misguided. But that was Boltzmann's program.

One of the most influential thinkers of the new school of thought was Ernst Mach. As much a philosopher as a physicist, Mach subjected the principles of explanation in science to the most intense scrutiny. He was a radical empiricist, and he held that only things actually experienced were objectively real. Everything else was our construct, and subject to doubt. He even considered the possibility that atoms existed in other than three dimensions.

Mach conceded the many successes of the atomic theory, but he resisted the implication that this proved atoms existed. Rather he regarded atoms as convenient fictions—possessed of the same degree of reality as, say, the Average Woman. Insurance companies knew all about such a "woman": her age, her weight, her life expectancy. But this did not mean she was to be found anywhere walking down the street. "Have you ever seen one?" Mach would grumble loudly from the audience as Boltzmann lectured on his atoms.

Boltzmann battled it out. He fought in lecture halls and in the pages of the scientific journals. Never did he shrink from the debate. Indeed, he seems to have enjoyed arguments, and on numerous occasions went out of his way to arrange them. Boltzmann was one of the great scientific battlers of all time— one contemporary account referred to him as "a most remarkable polemicist; much feared in congress." The criticisms he wrote of his opponents' positions were scathing. His struggle with chemist Wilhelm Ostwald, another of his critics, grew so intense that at one point even Mach thought things had gone too far, and attempted a reconciliation.

Boltzmann's personality exhibited an amazing mixture of toughness and even arrogance in these arguments, together with modesty and friendliness in his personal relationships. I am astonished by the separation he was able to maintain between these two spheres of his life. Boltzmann's published attacks on Ostwald are withering, but the two remained close personal friends throughout their rivalry. They often visited one another's homes. After Boltzmann's tragic death and the final resolution of the debate on the existence of atoms, Ostwald acknowledged his rival's success, and he wrote of him as "a man superior to us all in intelligence, and in the clarity of his science." Only a man of fundamental goodwill could have written such a graceful acknowledgment—and only one of goodwill could have received one from a man against whom he had written so scathingly.

Nevertheless, the attacks from all sides, and the persistent misunderstanding of his work, weighed on Boltzmann. A textbook of his day defined atoms as "imaginary units of which bodies are aggregates." So august an authority as the legendary eleventh edition of the *Encyclopaedia Britannica* refused to commit itself as to their reality. My general impression is that people were more inclined to accept the existence of atoms in Boltzmann's youth than toward the end of his life. The more time passed, the more he came to be regarded as a dinosaur, a throwback to an outmoded way of thought.

Shortly before Boltzmann's death, Einstein published a paper in which he argued that Brownian motion—an incessant jiggling of particles suspended in a liquid—resulted from the continual bombardment of these particles by atoms. Einstein's contribution turned out to be crucial, and within a few years it persuaded most people of the truth of the atomic hypothesis—and by implication, of the correctness of Boltzmann's mechanistic understanding of thermodynamics. But there is no evidence that Boltzmann ever learned of this work, and he did not live to see the vindication of his views.

"I am conscious of being only one individual struggling weakly against the stream of time," he wrote, and elsewhere:

I see myself as a man grown old on scientific experiences. Yes, I might even say that of those who embraced the old with a full heart I alone am left. . . . I present myself to you therefore as a reactionary, a man left behind.

During the closing months of his life, Boltzmann's mental illness rose up to claim him. These months were marked by near-unremitting despair. He was a speaker of great skill and his lectures were famous, but he continually dreaded that his memory would fail him one day in front of a class. He was subject to recurring anxieties, and confided to friends that he was losing his creativity, that his nerves were overstrained, and that he suffered from neurasthenia. Toward the end, his eyesight—never good—deteriorated still further; and he suffered from asthma, angina pectoris, and severe headaches.

"Father gets worse every day," his wife wrote to their daughter. "I have lost my confidence in the future. I had imagined a better life." In the fall of 1905 he decided to visit a mental hospital, but shortly after arriving, he left and returned home. He had announced a series of lectures for the following summer but, because of his condition, was forced to cancel them. That winter two students visited him at home in order to take an oral examination: as they left the house, they heard from the front hall a series of heartrending groans. Colleagues feared he would never again be able to exercise his professorship, and spoke of the need to keep him under constant medical surveillance.

In September of 1906 he took his family to Duino, a beautiful resort on the Adriatic. On the morning of his death he seemed unnaturally excited. While his wife and daughter went swimming, he hung himself in the hotel room. His daughter, whom he deeply loved, was the first to discover the suicide. "Never would I have believed such an end was possible," he had commented shortly before to a friend.

3

The Magician

George Gamow

Anyone spending an evening with George Gamow was assured of a good time. He himself loved a good time and he went out of his way to make sure he had one. He made up funny poems. He drew delightful, whimsical drawings. He organized charades and skits. He waylaid people, good-looking young women in particular, to demonstrate his latest magic tricks.

Gamow began his career as a nuclear physicist, switched in midlife to astrophysics and cosmology, and ended up working on the problem of the genetic code. Some of his colleagues resented him. Some felt he did not take science seriously enough—all this playfulness, this changing of fields. But he did take science seriously. He took it with all the seriousness of a game. George Gamow loved nothing so much as a good time; and science—all of it—was to him the best time of all.

Gamow delighted in telling stories. Here is his story of his own birth. The child was too big to pass through the birth canal, and the mother's life was despaired of. Eventually, the decision was made to cut the baby into manageable pieces—merely in order to save the mother's life, you understand: a comment on the state of medical knowledge in czarist Russia, if not on a perhaps excessive tendency to embellish a tale. The storyteller (Gamow) was saved by a neighbor who had heard that a famous Moscow surgeon was vacationing nearby. After being roused out of bed in the middle of the night, the doctor performed a cesarean section on the mother, while she lay on a table in the Gamows' book-lined study. That was in 1904, in Odessa.

Both of Gamow's parents were teachers. One of his cousins was an astronomer who went to Italy to study with Giovanni Schiaparelli, famous for his claim to have discovered canals on Mars. But the cousin spent his time with a nihilist group instead. Eventually, the group attempted the assassination of a high government official. The attempt failed and the conspirators were executed: the incident became immortalized in Leonid Andreyev's famous novella *The Seven Who Were Hanged.*

When Gamow was thirteen years old, the Russian revolution broke out. Schooling during these times was sporadic at best. Odessa was often bombarded by enemy warships, and one day while Gamow sat by a window reading a book on geometry, the window was shattered by the explosion of a nearby shell. Warring factions occupied the city. Typhus and cholera were rampant. The water supply was shut off. Food ran short, and in the countryside peasants hoarded what little they had.

Gamow tells in his delightful and engaging autobiography, *My World Line,* of the mathematician Igor Tamm, who one day made his way on foot into the countryside in an attempt to gather food. The story of Tamm's adventure has nothing to do with Gamow, but it is so marvelous I cannot resist quoting it:

> When he arrived in a neighboring village, at the period when Odessa was occupied by the Reds, and was negotiating with a villager as to how many chickens he could get for half a dozen silver spoons, the village was captured by one of the Makhno bands, who were roaming the country, harassing the Reds. Seeing his city clothes (or what was left of them), the capturers brought him to the Ataman, a bearded fellow in a tall black fur hat with machine-gun cartridge ribbons crossed on his broad chest and a couple of hand grenades hanging on his belt.
>
> "You son-of-a-bitch, you Communist agitator, undermining our Mother Ukraine! The punishment is death."
>
> "But no," answered Tamm. "I am a professor at the University of Odessa and have come here only to get some food."

"Rubbish!" retorted the leader. "What kind of professor are you?"

"I teach mathematics."

"Mathematics?" said the Ataman. "All right! Then give me an estimate of the error one makes by cutting off Maclaurin's series at the nth term. Do this, and you will go free. Fail, and you will be shot!"

Tamm could not believe his ears, since this problem belongs to a rather special branch of higher mathematics. With a shaking hand, and under the muzzle of the gun, he managed to work out the solution and handed it to the Ataman.

"Correct!" said the Ataman. "Now I see that you really are a professor. Go home!"

Tamm never found out who the mysterious mathematician was.

Early in his life, Gamow's interests tended toward science. He was precocious, and while his fellow students were making their way through more prosaic topics, he studied relativity and differential equations on his own. Nor did the budding researcher neglect to further the experimental sciences. At one point his father bought him a small microscope, with which he proposed to test the validity of the Russian Orthodox dogma. Having dashed home from Communion one day with a bit of transubstantiated bread tucked away in his cheek, he compared the bread to a similar crumb from his kitchen dipped in wine beforehand.

Gamow spent a year at the university in Odessa and then transferred to Leningrad. But amid all the uproar of the revolution, schooling was erratic. Furthermore, he found that he knew more about some subjects than his teachers. "It was somehow immaterial," he later related. "I was around the university, I was not in the university. I was registered. I was getting credits, but in most cases not attending lectures." He began work in experimental physics—primarily because experimentalists were given a convenient place to hang their coats. But he soon found his forte to be theory. He took an adviser, who assigned him a research problem. But Gamow found the problem sterile and uninteresting. His attention kept wandering.

A photograph of Gamow survives from those days. A group of students and faculty are arranged about a visiting scientist. Gamow is the only one not wearing a coat and tie. His face is large, his gaze intense—he seems to be glowering at the camera. Although he is standing toward the rear of the group, his image leaps off the page. A powerful presence.

In 1928 Gamow was given the chance to leave Russia and spend the summer at the university in Göttingen. It was a happy chance for someone from the relatively sleepy backwater of Russia: in those days, Göttingen was one of the world centers of physics. There, one bright sunny afternoon, sitting in the university library and leafing through the pages of a physics journal, Gamow happened across a paper by Ernest Rutherford—the same who had so cruelly humiliated Cecilia Payne while she was a student.

In his paper Rutherford was discussing the phenomenon of radioactivity, the emission of particles by unstable atomic nuclei. The point of Rutherford's paper was that by every known principle of physics, radioactivity ought never to occur. Nuclear constituents were known to be held within their nuclei by attractive forces, forces so powerful as to make escape quite impossible. An analogy might be that of a prisoner trapped within a cell.

Rutherford in his paper was proposing a means whereby such a prisoner might worm his way out under certain circumstances. The proposal, however, did not appeal to Gamow. It struck him as contrived and artificial. But even as he sat there, the correct explanation came to him. It came in an instant, fullblown, and within a matter of days he had worked out all the details.

It was the first major discovery of Gamow's career, and it was in many ways a remarkable triumph. His theory accounted for essentially every detail of the process of radioactivity, and this in spite of the fact that nuclear physics was still very much in its infancy in 1928. In those days people did not even know about the neutron, one of the two particles of which nuclei are constructed. Gamow's theory, in fact, was the first rigorous treatment of any nuclear process ever to be proposed. It transformed nuclear physics from its initial fumbling and imprecise state into a modern, quantitative field of study.

Most remarkable of all was the nature of Gamow's account of radioactivity. For in his paper, he makes not the slightest attempt to explain how the escaping particle performs its Houdini-like trick. Indeed, in his paper he is hardly concerned with particles at all. Instead, he is concerned with waves.

These were the new, radically incomprehensible waves of quantum theory. Quantum mechanics is one of the dominating theories of our age. There is hardly a single facet of physical science that it has not transformed. The theory was developed in a monumental effort extending over the first quarter of the century, culminating in the proposal by French physicist Louis de Broglie that to every subatomic particle there was associated a wave. Shortly thereafter, Austrian physicist Erwin Schrödinger wrote down an equation governing the propagation of these waves. With the formulation of this equation, the theory was essentially complete, and the debate came to a close.

But the debate on the *interpretation* of quantum mechanics has continued to this day. The theory is profoundly mysterious. It gives a detailed set of instructions for calculating subatomic processes, yet fails to provide the slightest comprehension of how these processes take place. Indeed, it seems to deny the very possibility that such a comprehension can ever be reached. So enigmatic are the theory's implications that Richard Feynman (chapter 6) once commented, "I think I can safely say that nobody understands quantum mechanics." Another physicist has put things more succinctly: "Quantum theory is magic."

The de Broglie waves that quantum mechanics invokes possess a curious combination of existence and nonexistence. They are there, and they are not there. On the one hand, they exhibit some degree of hard, physical reality in that they are influenced by external circumstances—they curl around obstacles, pass through holes in walls, and the like. But they cannot possibly represent a real physical vibration: among other things, their strengths are sometimes imaginary—the square root of a negative number. It is the subatomic particles that are real, not the waves that accompany them.

The connection between wave and particle is that the wave's square (reversed in sign if negative) gives the probability of finding the particle. Where the wave is big, there you are likely to find the particle; where the wave is small, you are not likely to find it. Gamow applied this prescription to escape from a radioactive atomic nucleus. The particle imprisoned within the "cell" has an associated wave that exhibits troughs and crests, and that implies that the particle can be located anywhere within the enclosure. The particle is forbidden to penetrate the solid rock comprising the walls of the cell. But the Schrödinger wave can and does penetrate these walls—although Gamow found that, rather than vibrating within this region, it smoothly decreases. Finally, in the region outside the prison, the wave resumes normal form, although diminished by its passage through the wall. This reduced strength corresponds to a small probability of finding the particle outside. But although small, the probability is not zero: there is a definite chance of the particle's escaping from the nucleus and flying away into space.

As worked out by Gamow, this is what quantum theory says of the process of radioactivity. This is *all* that it says. Notice the fundamentally incomplete nature of its account. In a very important sense, the theory entirely fails to describe what happened. After all—how did the prisoner do it? There was no crack in the walls through which he managed to worm his way. Nor was there some pliant guard about, available to be bribed. Nothing in the quantum-mechanical theory of radioactivity explains how a subatomic particle performs its magic and makes its way through impenetrable walls.

Most people would say that science is the effort to comprehend the natural world. But quantum mechanics has turned this view on its head. Rutherford sought to understand radioactivity by proposing a detailed mechanism whereby the process takes place, and in this effort he failed. Gamow succeeded, succeeded even as he sat reading Rutherford's paper, because he did *not* seek understanding. Rather, he was thinking only about the remarkable formalism of quantum mechanics. He was content with magic.

Gamow's schooling did not so much end as fizzle away. He lost interest in his Ph.D. thesis research and sent the University of Leningrad his work on radioactivity instead. It seems that they never got around to awarding him a degree.

As so often happens in such matters, he was not the only one to have thought of the penetration of barriers by quantum waves. Two Princeton University physicists, Edward Condon and Ronald Gurney, did much the same work and published it at much the same time. But news of the competition did not reach Russia, and when he returned home, Gamow found himself something of a hero. "A son of the working class has explained the tiniest piece of machinery in the world," extolled one newspaper, and *Pravda* went so far as to print a poem in his honor: ". . . Why, this working-class bumpkin, this dimwit, this Gyorgy Anton'ich called Geo., / He went and caught up with the atom and kicked it about like a pro . . ."

The son of the working class was given permission to travel abroad yet again. He spent time at Rutherford's laboratory in Cambridge and at an institute in Copenhagen. Upon returning home in 1931, though, Gamow found the political situation dramatically changed. Up to then, aware of the relatively backward state of Russia as compared to Europe, the government had taken pride in Russian scientists who were invited abroad. But now science had become just one more arena in which the battle against capitalism was to be waged. The era of state control of science had begun in earnest.

Everyone knows of the perversion of Russian biology by Lysenkoism. The field was crippled for decades by this intrusion of dialectical materialism into the laboratory. Physics, too, had its share of problems. For some strange reason, the quantum-mechanical uncertainty principle was decreed contrary to Marxist thought, and once while giving a public lecture, Gamow was forcibly interrupted by a government philosopher when he brought it up. The audience was sent off packing, and Gamow was instructed never to mention the uncertainty principle again. Einstein's theory of relativity, too, was decreed invalid. Over time these prohibitions were lifted, and by now some of the best work in physics is done in the former Soviet Union. But the pervasive influence of state philosophy persisted for decades.

When I was in graduate school and studying Einstein's theory, I once had occasion to consult a monograph on relativity written by a Russian physicist. The book was the typical maze of equations and abstruse symbols one expects of an advanced work in physics, and it was obvious that the author had not the slightest interest in Marxist philosophy. Nevertheless, the subject did come up—once. It was at the very close of the introduction, in a paragraph that stuck out like a sore thumb and four missing fingers. There the author explained that it was only through the writings of Lenin that he had come to a proper understanding of relativity, and only through the study of dialectical materialism that he reached a proper comprehension of the discoveries that he himself had made. This silly piece of nonsense had obviously been inserted at the last minute on the insistence of some government official.

Gamow found himself trapped in a deteriorating situation. He was denied permission to attend an important conference in Rome in 1931, and the intrusion of state censors became daily more restrictive. It was difficult for him to support himself. Pay scales were so low that he was forced to take five different jobs. Most of Gamow's time was spent waiting in line at cashiers' desks.

The rest of his time was spent planning an escape.

The story of Gamow's departure from Russia reads like a paperback thriller written by someone in an ironical frame of mind. Gamow and his wife pored over maps. Their attention was drawn by a stretch of the Black Sea leading from Crimea south to Turkey. They purchased a frail, collapsible canoe and hoarded food for months. They managed to wangle permission to vacation at a resort on the shores of the Black Sea. There one morning they paddled off, telling their fellow vacationers that they were planning a trip to a local observatory. But once out of sight of shore, they turned due south and commenced paddling across the 170 miles of open sea.

Forty-eight hours later they ended up not far from where they had started: cramped, cold, hungry, and exhausted. They had been thwarted by a windstorm and had narrowly escaped with their lives. Once back home in Leningrad, they turned their attention to the border with Norway. An oceanographic laboratory had been set up not far from Murmansk. There they made several forays to scout the chances of a second attempt. But the Soviet navy was there in force. They returned home once more.

A few days later, Gamow received a letter officially *instructing* him to leave the country. He was to attend a conference in Brussels: his passport, visas, and rail ticket were to be provided by the Soviet government. That was in the fall of 1933. Gamow and his wife left for Brussels and never went back.

The only item of note that occurred at the conference was a major revolt during his talk, in which both French and non-French physicists demanded that he stop trying to speak French. Within a year, he had gotten a job at George Washington University in Washington, D.C. Upon hearing of his appointment, he bought a ticket to Seattle—he had never been in America before.

Gamow's work is marked by a remarkable inventiveness and originality. In each of the disciplines in which he worked, he imposed his own special stamp. Most scientific papers break little new ground and are chiefly concerned with cleaning up this or that messy detail. But in Gamow's writings, one finds insight upon insight.

His papers are a pleasure to read. While many are heavily technical, others contain not a single equation or mathematical symbol. These are unusually short—often a mere three or four paragraphs—and they are marked by an intense compression of thought. Every sentence says something new. It is a style much in favor among scientists, who could never be accused of garrulity in their scientific communications. But my own impression is that Gamow developed the style to an unusual degree. He published papers in German, French, and English, but never, curiously enough, in his native Russian.

Gamow's technical abilities were great, but he had little interest in technique for its own sake. He was happier when establishing the broad outlines of a subject than when working out all the details. He preferred fields that were new and unexplored, in which surprises awaited the traveler at every turn. It was this predilection that in Göttingen had led him to apply quantum theory to nuclei, while everybody else was applying it to atoms and molecules and raking in Nobel Prizes in the process. But by the mid-1930s, nuclear physics, which he had helped establish, was becoming too well established for his liking. He moved into astrophysics.

There his most significant contribution dealt with Einstein's big bang theory of the origin of the universe. It was a contribution that, as usual, carried with it Gamow's own imprint, and it radically altered the very nature of the field.

The mathematics in which Einstein's theory is couched is extraordinarily abstruse, and it is unusually difficult to work with. The theory is so technically complex that many of its most significant features remained undiscovered for decades. To this day, after more than three quarters of a century of effort, research continues on the task of discovering just what Einstein's theory predicts. This work is mathematical, abstract, and far removed from physical reality. At the same time, dealing as it does with the nature of space and time, with the expansion of the cosmos and its overall geometry, and with the very creation of the universe, the big bang theory generated much work that is purely philosophical in tone. This work, too, is abstract and far removed from reality.

But Gamow avoided all the mathematical complexities and the abstract philosophical implications. The approach he took could almost be termed pedestrian, had it not turned out to be so fruitful. He treated the big bang as a physical event that had actually occurred—an extraordinary event to be sure, but one not so very different from any other. He investigated that event's consequences. He asked if any of these consequences still remained. In this he was, if you will, doing a sort of cosmic archaeology: looking about for traces of our past.

The Milky Way galaxy in which the Earth is situated is a stupendous, disk-shaped aggregation of maybe 100 billion stars. The Milky Way, easily visible on clear moonless nights, is that disk seen edge-on. Also visible to the naked eye, though with greater difficulty, are a few dozen other galaxies; astronomical photographs reveal myriads more. So far as we can tell, the space between galaxies is largely empty: they seem to be the fundamental units into which moons, stars, and planets are organized.

Galaxies are regions into which the material of the universe has been concentrated. They are knots, condensations in its overall structure. In collaboration with physicist Edward Teller, Gamow asked how these condensations might have formed. According to Einstein's scenario, the universe originated in a dense, superheated big bang and then began expanding and cooling. It occurred to Gamow and Teller that in such an expansion a stage must have been reached in which the universe as a whole was just as dense as galaxies are today. They argued that this epoch most likely marked the point at which these structures formed. Gamow and Teller identified various processes that might have intervened at this stage to cause the expanding debris to coagulate into a multitude of subunits, later to become galaxies.

The scenario they proposed forms the basis for present-day thinking: galaxies are relics of the big bang. In related work, Gamow and his students asked whether nuclear reactions in the early universe might have synthesized all the chemical elements we find today, and whether the blaze of light accompanying its creation might still remain and be detected.

George Gamow was exuberant, jovial, rough, and friendly. He stood 6 feet 6 inches tall; skinny in his youth, he weighed 250 pounds toward the end of his life. (Remarkably enough for such a huge person, his voice was unusually high-pitched.) He had a large face, prominent ears, and intensely blond hair. His personality, too, was immense, and he was the sort of person who seemed to fill the room. His memory seemed limitless, and he would quote poetry by the hour.

Gamow's capacity for alcohol was equally limitless. He drank incessantly and hugely. Something of a mystery surrounds this aspect of his life. It is widely believed that in his last days his mind was poisoned by alcoholism, but I have been told by several colleagues who knew him personally that the belief is false and that no matter how much he drank he never lost control. At any rate, it is clear that Gamow's last years were spent in an almost perpetual state of illness, many of his problems apparently related to his lifelong consumption of alcohol. He died at the relatively early age of sixty-four.

Fascinated by magic tricks, Gamow worked hard to perfect them. He loved nothing so much as a good joke. He would organize skits in which prominent physicists made thinly disguised appearances. He wrote a parody of the book of Genesis in which the theories of a colleague were lampooned. He built little Rube Goldberg gadgets, replete with drawings and legends in several languages, in which grand scientific principles were illustrated. These things would never actually work—they were made of plywood and string: they could not possibly work. But working was not the point. The point was the fun of it all.

Gamow began one technical paper with a quote from *Alice's Adventures in Wonderland*. His notebooks and letters to colleagues were filled with silly rhymes and playful drawings. They were written in a handwriting that was huge, chaotic, and sprawling, and they bounced madly from subject to subject. His spelling was an amazement.

Unlike many scientists, Gamow wrote for the public, and he did so prolifically. His books are classics of their genre, and they move from mathematics to physics to astronomy to biology with gracious ease; this from a man who did not speak English until he was in his twenties. Perhaps his most famous writings are the Mr. Tompkins stories. Mr. Tompkins (his full name was C. G. H. Tompkins, "c" being the mathematical symbol for the speed of light, "G" for Newton's constant of gravitation, and "h" Planck's constant—more fooling around) was a bank teller whose father-in-law was a professor of physics. Tompkins had a habit of nodding off at the old man's lectures and falling into a dream, and in these dreams he would encounter the very things the professor was explaining. In one book the speed of light was reduced to a mere few miles per hour, and while driving in a car, Tompkins would directly experience relativity's many strange phenomena. The dreams are a wonderful literary device, and they render the most abstruse of subjects easily and vividly comprehensible.

Generations of physicists, myself included, recall with fondness and delight reading Gamow's books as youngsters. They have the same easy, cheerful style that animated his conversation. Some books explaining science for the public strike a tone of high seriousness, not to say mysticism; but Gamow's style was that of the storyteller, the man who has just learned something wonderful, something fascinating, and who cannot wait to tell it to you.

In the summer of 1953, a friend walked into his office and showed *him* something wonderful. It was Watson and Crick's just-published paper on the structure of DNA, and the event marked yet another transition in Gamow's career: out of astrophysics and into biochemistry. He was nearly fifty years old at the time.

Gamow's curiosity was aroused by the problem of the genetic code. By 1953 it was known that the nature of every organism was controlled by the blueprint contained within its DNA. But it was also known that this control was not exerted directly. There was an intermediate step: the synthesis of a set of enzymes, which migrated outward from the DNA to catalyze the biochemical reactions taking place throughout the cell. The problem of the genetic code was the problem of how the blueprint contained in the DNA carried over into the second blueprint contained in the enzymes.

Enzymes were long chains composed of amino acids. There were twenty different sorts of these acids. The famous achievement of Watson and Crick was the unraveling of the analogous structure of DNA. Each strand of their double helix was a chain of nucleotides, of which there were but four: adenine, thymine, guanine, and cytosine. The patterns running down the two strands differed, but they were not independent. There was a curious rule of association: wherever adenine appeared on one strand, thymine appeared across from it on the other; and wherever cytosine appeared on one, guanine appeared on the other.

Biochemistry is a subject of fearsome complexity. But Gamow in his original paper on the genetic code paid not the slightest attention to any of its details. Rather, he concentrated on its most general, abstract features. In that paper he described DNA as "a long number written in a four-digital system," referring to the four nucleotides of which it was composed; and similarly, he wrote of enzymes with their twenty amino acids as "long 'words' based on a 20-letter alphabet." In so doing, he was putting his own peculiarly individual stamp on the field. The very language in which he was speaking was uniquely his own.

In Gamow's language the problem of the genetic code was one of translation. How could something written in a four-symbol code be translated into a twenty-symbol code? Put this way, the question could be tackled entirely apart from the many details of biochemistry—details that in Gamow's view just got in the way. To him the question was utterly abstract, and the tack he took in looking for a solution was similarly abstract.

He tried some guesses for the nature of the translation system. The simplest would be that each "number" along the DNA strand determined a corresponding "letter" along the enzyme strand. But this would not do, for since there were only four separate numbers, enzymes would be composed of only four separate letters. A second possibility was that each *pair* of numbers determined each letter, but this would allow for only sixteen amino acids. Finally, a system in which letters were determined by triplets of numbers turned out to allow not too few amino acids, but too many.

Gamow's final guess was ingenious. He considered four particular nucleotides—two of them sitting at a particular location on one strand of DNA and two on the other, the locations chosen to form the pattern of a diamond. He figured that the hole at the center of each diamond would have a *shape* that depended on just which nucleotides were present. His proposal was that each amino acid would be coded for by that shape. Maybe the amino acids actually got stuck in the holes, and then were strung together like beads on a string to form the enzyme. At any rate, because one of the nucleotides of the diamond was not random, but was determined through the rule of association by those sitting across from it, there turned out to be *exactly twenty different possible diamonds*. And twenty was just the number of different amino acids.

The delightful thing about Gamow's proposal is how it unites what appear to be unrelated, disparate elements. Twenty amino acids, four nucleotides, and a peculiar rule of association: suddenly everything falls into place, and by an inescapable logic a four-digital system translates into a twenty-letter alphabet. It is a proposal of simplicity, elegance, and beauty—often the hallmark of a good explanation.

Unfortunately for the storyteller's art, Gamow's proposal turned out to be wrong. It required more than a decade of effort by biochemists the world over to find the right answer. The true genetic code is complex, and it involves many of the messy details that Gamow had hoped to avoid. To me, at least, the true code is less elegant, less beautiful than his proposal. And I feel confident that he was led to his proposal by just such aesthetic considerations. For in theoretical physics, the explanation that is elegant, that has the greatest degree of mathematical beauty, often turns out to be correct. But this seems not to be the case in biology.

In an age of specialization, Gamow was a generalist. All of science was his province. Indeed, the very style of his work kept changing. He made no attempt to come to what we normally think of as an intuitive understanding of radioactivity, contenting himself rather with developing a theory based on the magical formalism of quantum mechanics. His work on cosmology, on the other hand, was wholly an effort to develop a physical comprehension; avoiding philosophical speculation and mathematical formalism, he worked to comprehend the big bang in all its details. In his biological studies, his role was that of the puzzle solver—one of his popular books had been on puzzles—and he focused attention on the abstract, mathematical aspects of the problem of the genetic code.

Common to all of Gamow's work are the qualities of playfulness and inventiveness, and a resolute refusal to be trapped within a ponderous consistency. The gleam I imagine in his eye as he worked out the solution to a scientific problem is the same gleam I see as he recited the poetry of Pushkin by the hour, or demonstrated to an impromptu audience some new and startling trick of magic. Toward the end of his life, recalling his perilous flight across the Black Sea in a frail canoe, a flight in which he had been in danger for his life, one particular image stood out in his memory: that of a porpoise he had glimpsed, momentarily suspended in a wave illuminated by the setting sun. How Gamow-like!

4

A Gentleman of the Old School

Homi Bhabha and the Development
of Science in India

Homi Bhabha was not in the habit of traveling alone. When he drove out to the Bombay airport one day on one of his many trips overseas, he took along a retinue: aides, associates, secretaries. The day was Sunday, but a few last-minute instructions might have occurred to him before boarding the plane.

Trained as a physicist, Bhabha was an administrator of astonishing ability, capable of pushing to completion the most gigantic, most visionary of projects. He was founder and director of one of India's most prestigious research institutes, creator of its atomic energy program, secretary to the government of India, chairman of numerous international agencies. It can be argued that a good chunk of all modern science and technology in India owes its origin to this one man. His research work, on the other hand, was deeply abstract. His knowledge of mathematics was extensive, even compared to that of other physicists. In style his work was formal, refined, elegant—almost Mozartean in tone. I know of no other instance in which two such radically differing skills were combined in the same person.

A man who dined with Bhabha on one occasion has reported to me that Bhabha had brought along an aide whose sole function appeared to be to pay the bills. It would not have been seemly for a gentleman to be seen dipping his hand into his pocket, and above all else Bhabha was a gentleman. Born in India to a wealthy family in 1909, he was aristocratic, international in outlook, and a man of exquisite refinement and taste. In every major capital of the world, Homi Bhabha had his favorite hotel. He knew who was singing at La Scala. He knew what was showing at the Tate.

Bhabha's education was select, and Western in style. His father was a barrister, one grandfather a government official and a Companion of the Order of the Indian Empire, another a well-known philanthropist. His family was related by marriage to the multimillionaire Tata family of industrialists. In the Tata household across the street, Bhabha would hear as a boy discussions of giant, long-term projects relating to the industrial development of India. Back home in his father's panelled study was a remarkable collection of books and recordings of Western classical music. He and his friends would listen to these records in hushed, reverent silence. A photograph of Bhabha in these days shows a teenager dressed in dark suit and tie, reclining in a formal pose in an overstuffed armchair. In his lap is a book on El Greco.

It was inevitable that such a youth would be sent to England for his advanced studies. Bhabha went to Cambridge University—Cecilia Payne's old school—and he attended a college that had recently received a huge endowment from an uncle. His father wanted him to go into mechanical engineering, but Homi wanted to do math. A compromise was reached in which his father agreed to finance Homi's mathematical studies if he started by getting a solid grounding in engineering. Homi got his First in engineering in 1930 and then went on to get a second First in mathematics one year later.

Through college he cut a wide swath. A fellow student who went on to a highly distinguished career at Cambridge has frankly confessed to me that he had felt positively over-awed by his friend in those days. Bhabha collected prize after prize for his mathematical abilities. He was an accomplished artist (his paintings and drawings have since been published). He designed the stage sets for a Calderón play and for a Mozart opera.

He studied harmony and counterpoint, and he wrote a symphony in the classical style. He showed up at a department party dressed as a Spanish grandee. He wrote a skit:

SHAKESPEARE ORDERS LUNCH

SHAKESPEARE: What ho, without!
SERVANT: My lord.
SHAKESPEARE: Full twice or thrice
              Have I, with lusty and barbated speech
              Sought to affront the portals of thine ears.
SERVANT: Pardon, sweet lord.
SHAKESPEARE: 'Tis granted. Look you now,
              The time approaches when my corporal
              frame
              For lack of food grows incorroborate;
              Fetch me my specs, that I may make perusal
              Of whatso'er of viandry is set
              For our engorgement. . . .

After getting his Ph.D., Bhabha remained at Cambridge in a research position. He worked on cosmic rays. High-speed subatomic particles from space, they were the most energetic objects known in those days—and they still are. Their tremendous energies greatly exceed those produced in the largest particle accelerators. Even had the mammouth Superconducting Super Collider been constructed, it would have been incapable of matching them. The origin of cosmic rays is poorly understood—they come from somewhere far out in space—but even in the 1930s it was clear that they must have been underway for thousands, possibly millions of years before reaching us.

Remarkably they appeared to arrive, not singly as individual particles, but in sudden showers in which great numbers were detected at once. It was almost as if they had been produced and traveled through space in groups. But this could not be, for the showers were known to consist of particles of differing velocity. In transit the more rapid particles would have outstripped the slower, elongating the hypothetical group and failing to produce the observed brief shower.

Things became clearer when a series of experiments actually succeeded in documenting the process. In perhaps the prettiest, a horizontal metallic sheet was inserted into a cloud chamber, the chamber recording the passage of each ray as a track. A photograph would reveal an individual track impinging upon the sheet of metal from above—but large numbers of tracks fanning outward beneath it in an inverted cone. Within the sheet, each particle was multiplying into a spray. Presumably the same process was going on in the upper atmosphere. A picture began to take shape in people's minds of individual cosmic ray particles flying for uncounted ages through interstellar space, each one ultimately striking the Earth's atmosphere and spawning a group that arrived at the ground as a shower.

But how did the particles spawn the groups? It was as if a flying golf ball, upon striking the grass, spontaneously transformed into a collection of golf balls. Bhabha took the point of view that the cloud chamber photographs were providing explicit documentation of Einstein's famous relation between mass and energy, $E=MC^2$. The energy of motion of the impinging particle was turning into mass—mass in the form of other particles. A proper treatment would require a union of Einstein's theory with quantum mechanics, the theory of subatomic processes. This union had recently been achieved by Cambridge physicist P. A. M. Dirac. Bhabha and a colleague worked out a complete theory of the production of cosmic ray showers according to Dirac's theory.

The picture they evolved was that of a rapid and escalating back-and-forth interchange between matter and energy. The process began when a single cosmic ray particle collided high above the surface of the Earth with a molecule in the atmosphere. In the collision a photon would be produced—pure energy: a minute burst of light. The photon, now accompanying the deflected cosmic ray, soon itself interacted, transforming into a pair of particles. The single incident cosmic ray particle had now become three. Each of these subsequently collided with another air molecule, producing its own photon, and the process continued in a snowballing cascade. Thus the shower.

Dirac's theory is aesthetically exquisite, but it is couched in a complex mathematical language, and the calculations it necessitates are daunting. Since the 1930s various shortcuts have been found, smart tricks that obviate the need for some of the harder slogging. But these were not available in those days. People used to lay out oversized sheets of wrapping paper on tables, and fill them with their calculations.

The urbane, charming Bhabha could not have chosen a more unlikely mentor than Dirac with whom to work. By all accounts Dirac was an austere, otherworldly individual. He was an uncommunicative soul, and hardly ever told anybody what he was working on. His theory involved the mathematical symbol <l l>, called a braket, of which the two halves <l and l> also held significance. Dirac called them "bra" and "ket" respectively, and it seems that he was utterly astonished to learn one day that the first of these terms also carried another meaning. (He was married at the time.) The story is told that a colleague once tried to give him a book with which to while away the hours on a long train journey. But Dirac refused the gift, explaining that reading "interfered with thought."

The cascade theory of cosmic ray showers was wonderfully successful. It was almost *too* successful. It accounted beautifully for their observed properties, but it failed to account for one annoying fact—not all cosmic rays made showers. Indeed, there was known to exist an anomalous component of cosmic rays that could penetrate great distances, even through solid lead, without multiplying into a shower. According to Bhabha's calculations, such a thing was entirely impossible. What to do? For a time no one was quite sure: the possibility was even bruited about that Dirac's theory was wrong in some essential respect.

But eventually Bhabha settled upon a different point of view. He proposed instead that high in the atmosphere, an elementary particle unknown to physics was being produced in cosmic ray collisions. He showed that if this new particle possessed a particular mass, roughly midway between that of the electron and the proton, it would fail to produce a cascade. Subsequent experiments succeeded in confirming his guess by directly observing the hypothetical particle. People named it the muon—it will figure strongly in chapter 7 of this book.

Remarkably enough, however, Bhabha's success only deepened the mystery. How could these experiments have found the muons? His original idea had been that his new particle was a more or less permanent entity. But it was not: muons were soon discovered to be unstable. Laboratory experiments showed them to fragment into other particles quite rapidly. So rapid were these decays that the muons could not have traveled far from their point of creation in the upper atmosphere before disintegrating. In particular, they could not possibly have reached the ground. But they were reaching the ground: obstinately, muons persisted in showing up in ground-level detectors.

Bhabha's response to this conundrum was marvelous. He decided to think of the cosmic ray muon in a new and richly creative way. He thought of it not as a previously undiscovered elementary particle, but as a tiny, rapidly moving clock. Each tick of such a clock carried it closer to the moment of its decay. But according to Einstein's theory of relativity, clocks in motion ticked more slowly than those at rest. So a rapidly moving muon would live longer than one that ambled along more slowly.

Einstein's proposal had been astonishing when he made it, just after the turn of the century. It is still astonishing. His argument is one of deep beauty, profound in its implications and breathtaking in its scope—perhaps the most perfect instance I have ever seen of the power of abstract reasoning. But no familiarity, no study, has ever accustomed me to the idea that time can slow.

The slowing-down of time is not great for clocks that travel slowly. At the relatively low speed of an automobile, or even of a spacecraft, nothing much happens to time. But according to Einstein, remarkable things would happen if objects traveled at velocities close to that of light. These would exist in a kind of suspended animation: lugubriously ticking, each second requiring hours, each hour months . . . a curious combination of blazing speed and eerie stillness.

Bhabha recognized that cosmic ray muons constituted just such high-speed "clocks," and that if relativity was correct, their rate of ticking would be slowed in such a manner. In this way he accounted for their anomalous survival during their passage through the atmosphere down to the ground. It was, indeed, a double triumph, for when Einstein formulated his theory, there had not been the slightest possibility of testing it experimentally. To do so would have necessitated accelerating some clock to impractical velocities. For decades, then, the notion of the slowing-down of time remained in a kind of limbo; widely accepted, yet unproven. Bhabha's insight provided the first experimental test of Einstein's prediction.

In 1939 Bhabha returned to India on vacation. While he was there, the Second World War broke out. After thirteen years at one of the world's centers of research, he suddenly found himself stranded in a relative backwater.

He spent the war at an institute in Bangalore. There his work, in the assessment of his colleague and friend M. G. K. Menon, altered in character. It grew more abstract. Bhabha found himself growing less and less concerned with application to observed phenomena, and steadily more interested in questions of internal coherence and formal elegance. In later life Bhabha commented that this work, though it earned him less recognition, had given him in the long run the deepest satisfaction.

And it was during this time that another change occurred in Bhabha, one that I find quite amazing. While at Cambridge he had given every indication of intending to remain there forever. Nothing I have learned gives the slightest hint of any desire in him to return home during these days. And indeed, by the end of his exile during the war he had received a number of job offers from the West. But he rejected them all. For the rest of his life Bhabha never accepted a permanent position abroad.

By the war's end Bhabha, formerly more British than the British, had become Bhabha the patriot, Bhabha the citizen of a new nation. And it is significant that his decision to remain in India went hand in hand with a decision to create a new scientific institution there. Bhabha set forth upon a program of building up science in his home country. He did not accept a job at some research center—he built his own. In a letter to one of the Tata family's charitable trusts, he proposed the creation of a new institute for fundamental research.

It is difficult for a Westerner to appreciate the context in which this decision took place. India at the time boasted of a number of world-class scientists who had been honored by the Nobel Prize or membership in the Royal Society, and the institute at which Bhabha spent the war was a fine one. But these did not add up to a national environment conducive to the practice of science. Indeed, to an enormous degree Indians regarded their country as inferior in matters pertaining to modern development. As a matter of course, most looked to the West for inspiration. Bhabha was not alone in having gone overseas to college: endowed professorships at major universities often actually required the holders to have been educated abroad. So, too, with the Indian Civil Service, which insisted that its members be trained in England. Even after independence the presumption of European superiority remained—the first science adviser to the prime minister of India was British. And to this day Indian scientists prefer to publish in Western, not Indian, journals.

In such a context, the decision to create a major research center carried a particularly charged significance. What can account for Bhabha's change of heart during the war years? In a letter to a relative he wrote of his duty to stay in India, but I find this explanation hard to credit. His actions do not strike me as those of a person acting from a grim sense of duty. Rather, they strike me as the vigorous, happy actions of one who has suddenly found himself.

Perhaps when he returned home, Bhabha found within himself an aptitude for institution-building, for administration on a grand scale. Such an aptitude never could have flourished in a country like England, whose scientific institutions were already well in place, but it was free to flower in the relative vacuum of India. Perhaps, too, he found himself swept up in the exhilaration of a great moment. He had been too young when he left India to have taken part in the independence movement. But by the time he returned, it was obvious that independence would not be long in coming. It must have been a heady time, a time in which everything seemed suddenly possible.

V. S. Naipaul, born in Trinidad of Indian ancestry, has written of his childhood,

> I felt [then] that the physical conditions of our life, of-ten poor conditions, only told half the story: that the remnants of the old civilization we possessed gave the in-between colonial generations a second scheme of rever-ences and ambitions, and that this equipped us for the outside world better than might have seemed likely.

The Tata Trust responded favorably to Bhabha's proposal for the creation of a new research center. In 1945 the Tata Institute of Fundamental Research was opened. In the early days it was housed in an elegant mansion owned by Bhabha's aunt. She re-mained in residence, occupying half the house, and she moth-ered the scientists shamelessly, serving Homi tea in her best china, allowing the staff use of her kitchen and servants, and laying on a grand feast on Parsi New Year's Day. Bhabha's office was the room in which he had been born.

Within a few years the institute outgrew these quarters. It moved to the former Royal Bombay Yacht Club, a fine old Vic-torian pile on the waterfront complete with turrets, porches, and gardens. In the days of British rule no Indian had been al-lowed inside in other than a menial capacity: now the staff was housed in what had been the servants' quarters, and the library occupied what had been a formal ballroom. Funding for the in-stitute grew at an extraordinary rate, increasing by 30 percent per year during the first ten years. By the mid-1960s it em-ployed twelve hundred people and had moved yet again.

The institute's funding has come from a combination of pri-vate and public sources. But remarkably enough, one potential donor has been conspicuous only by its absence: the University of Bombay. One might have thought the university, an enor-mous institution located a mere few miles from Bhabha's, would have jumped at the chance to encourage a strong research presence on its very doorstep. But it did not. Neither the University of Bombay, nor any other university, has made a significant long-term contribution to the Tata Institute for Fundamental Research.

Throughout India the development of science and technology is a process that has taken place largely outside the university framework. With rare exceptions, conditions at Indian universities are inimical to the practice of science. During my visit to India in preparation for this chapter, I visited a major, publicly funded state university. I was shocked by the atmosphere of decay I found there. The physical plant was utterly demoralizing: paint peeling from the walls, corridors filthy, office windows broken, an inner courtyard a weed-infested vacant lot. As for the faculty, most did not even pretend to keep up with advances in their field, let alone participate in research. Instead, they were comfortably coasting downhill toward retirement. Those few who did engage actively in their professions did so in nearly complete isolation, and with the most primitive of equipment. Technical help was essentially nonexistent: one man with whom I spoke referred to the technicians as bureaucrats whose only function was to ensure that work orders be accompanied by the right paperwork, but who were completely incapable of doing anything else. As for the laboratory equipment, it would have made a high school kid from the West smile condescendingly. I was awed—and shamed—by the drive, the intensity of commitment, that enabled these people to conduct research in the midst of such appalling conditions.

The Tata Institute of Fundamental Research, in contrast, is internationally recognized as a center of excellence. India's first computer was built there; its first nuclear reactor would never have been built without help from the institute scientists. People there work in conditions an Indian university professor would find positively sybaritic, and its research program is competitive with that of any Western institution. The institute occupies a sleek, modern complex of buildings set amid perfectly tended formal gardens. Paintings adorn the walls, sculpture the entrance foyer. From office windows, the white towers of the Bombay skyline rise across the bay.

Walking the corridors there one day, it occurred to me that Bhabha's choice of architectural style for his institute might be significant. Not only in terms of research would the Tata Institute fit in well at MIT; its buildings would look at home there, too. Why had Bhabha adopted so resolutely Western an architectural tone? Why didn't the buildings there *look* Indian?

On the very day I arrived in India, the newspaper carried an article describing an address given before a science congress by the prime minister. In this speech, he pressed the delegates to cease doing what he called "Western-oriented science," and return to a more Indian mode of research. Was he right? Is science an exclusively Western phenomenon? It is difficult to express in words the sense of cultural dislocation, of an absolute and fundamental *alienness,* that I felt throughout my stay in India. I felt that I had entered a different universe. But at the Tata Institute, I felt the ease and comfort of one who has returned home. For all his claims to be developing indigenous technological expertise, was Bhabha's choice of architecture a clue to his true intent? In reality, was he merely attempting to impose foreign ideals onto an uncongenial environment?

There is in India today a group of people who would agree with the prime minister. They regard science as a Western import, and an unwelcome one at that. My guess is that these people are primarily motivated by sentiments of nationalism—by the wish that their country would abandon modern technological development and return to its ancient glories. In America as well, one can point to people who would agree with this sentiment. There are those who distinguish between the wisdom that is offered by science and the superior Wisdom offered by Eastern religions. And there are those who regard science as nothing more than a form of political domination.

None of these people would get any comfort from those with whom I spoke in India, however. During my stay, I raised the prime minister's speech with every person I interviewed—a mixture of scientists and foundation executives. Without exception, each one of them rejected his distinction utterly. There was no such thing as Western science or Eastern science, they argued, but only science. "I never regarded myself as learning the methods of the West when I was a student," one scientist told me. "I was simply learning my trade."

The prime minister's speech may have amounted to nothing more than a concession to popular sentiment. But my guess is that his point may indeed have a certain validity—but not with regard to the nature of science. I think his point is an important one with regard to the role that science plays in Indian society.

The tourist who remains in India's cities will return home with an utterly distorted image of this country. Three out of four Indians live not in these modern, industrialized centers, but in tiny farming villages. In these villages, few homes have running water or electricity. The predominant means of transportation there is the bullock-cart, and the way of life has hardly altered in centuries. The gap between the inhabitants of such a village and an Indian scientist is gigantic—overwhelmingly greater than that between this scientist and his Western counterparts. To the villagers of India, the activities of a scientist must appear as incomprehensible as those of a Martian.

Bhabha argued for the funding of fundamental research in India by invoking the "trickle-down" theory of development. He argued that the discoveries of research ultimately would emerge as benefits to the society as a whole. This happens in the West all the time—today's great scientific discovery becomes tomorrow's gadget, obtainable for a pittance at Kmart. But while these benefits do indeed accrue to India's middle- and upper-class population, they never touch its villagers.

For these people, the fruits of technological advance are of a wholly different sort—and they have nothing to do with the Tata Institute for Fundamental Research. The most impressive example I know of is the Satellite Instruction Television Experiment. Under this so-called SITE program, the government gave each village a television set, and educational programs were beamed nationwide via a geosynchronous satellite. Care was taken that these programs concerned themselves with technology appropriate to village life: crop rotation, water purification, and the like. Programs we might watch on *Nova* were sedulously avoided. The antennas by which these satellite broadcasts were received were themselves a triumph of appropriate technology—a mere few wires dangling from poles stuck in the ground. The SITE experiment reached twenty thousand villages and it went on for a year; its place has since been taken by a series of half-hour broadcasts each night. This modest, low-technology enterprise has done more to benefit the bulk of India's population than any amount of "Western-style" science.

In saying this, I am most definitely not arguing that science should be abandoned in countries like India. I am arguing that its justification has nothing to do with the trickling-down of technology; rather, the justification must involve arguments of a different sort. Indeed, I would say that one very good argument for science—science in India just as well as in the West— has nothing at all to do with technological advancement. Science is worth doing because it teaches us something of the true scheme of nature, and of our place in that scheme. It teaches us our address in the universe.

These are enough for me. They are enough for most scientists: I have yet to meet a one who works for the good of humanity.

"Homi was a distant alp. One did not presume to climb it."

In conversation after conversation with those who knew Bhabha, I heard in people's voices appreciation and respect for him—but I never heard friendship. Bhabha did not have personal friends at the Tata Institute. Throughout his career, there was always an element of social distance between him and the people with whom he worked. I think it is fair to say that in some fundamental sense Bhabha did not understand the life of the ordinary Indian. His wealth, his aristocratic upbringing, set him apart. In designing the institute, he took care to get an espresso machine for the cafeteria, but he allowed construction of staff housing to languish for years while working on the plans: this in Bombay, where real estate prices are among the highest in the world. He was not one to hang around the corridors chatting—most people there hardly ever saw him. There was a fastidiousness about him, a formality, that forbade casual contact.

Bhabha circulated at one point a memo to the institute staff dealing with its buildings and grounds. It makes for interesting reading:

> . . . Each member of the staff should have a sense of personal pride in these buildings, which have been given for his use, and it is his duty to take personal interest in their proper maintenance and to see that he himself uses them in such a way as to maintain their quality and cleanliness. . . . A member, who sees another not using the buildings properly should draw his attention to the proper conduct in such matters. If any person continues to misbehave, the matter should be reported to his superior.

This is not a memo to grown men and women. It is a memo from a father to his children.

The story is told that a group of scientists, recently arrived in India for a conference, found themselves transported about the country by bus while Bhabha himself traveled separately in his own private limousine. One of the conferees is said to have been so disgusted that he left the next day. On another occasion, Bhabha escorted a visiting lecturer to the hall where the visitor was to deliver his talk. They found every seat taken. Bhabha gestured to the people sitting in the front row, each one of whom rose and vanished, leaving Bhabha to attend to the lecture in solitary splendor. The first of these stories has been disputed, and I cannot vouch for its accuracy. But he does seem to have been the sort of person about whom such tales accumulate.

With only one exception, every photograph I have ever seen of Bhabha shows him dressed in a Western suit (the exception is the inauguration of India's first nuclear reactor, for which he wore national dress). His outlook was international, not Indian. My guess is that most of the people he regarded as his equals lived not in India but in the West. Each was an acknowledged expert in his or her field, and Bhabha was continually traveling abroad to drop in on them. There he would learn of the latest advances, and upon returning home, he would fill everyone in on the news.

Nevertheless, by all accounts Bhabha was an extraordinarily invigorating person to be around. Everything seemed possible in his presence. His ability to secure funding was legendary. Roadblocks that would have defeated another were surmounted through his personal connections with ease. Some new and marvelous project was always in the offing. At one point he brought in a biologist to head up a new effort in that direction (not bothering to tell anybody else at the institute about his new plans until after the position had been offered). Upon arriving, the biologist had a conference with Bhabha in which he was asked how much space the program would require. "Oh, maybe 5,000 square feet," he replied. "Nonsense," Bhabha responded. "You'll run out of that much space in no time at all." So he built him a whole new building.

In matters of research, Bhabha scrupulously left his scientists alone—but when it came to the institute's physical layout, he snooped about endlessly. He pestered the firm of architects that designed it over the tiniest of details. He worried about the location of the toilets. He had them knock down half-completed walls if their location did not suit him. When he got word that a rare tree was scheduled to be cut down at a nearby botanical garden, he arranged for the thing to be dug up and transplanted to the institute grounds.

Bhabha was a great patron of the arts, and through his wealth, he amassed a fine collection. Several well-known Indian artists had in fact been discovered by him. Most people who collect art hide it away: Bhabha set his collection out for view in the corridors and offices of the Tata Institute. Upon purchasing a new painting, he would experiment with various locations; hanging it one place and then another, asking the advice of friends. He did not merely want the institute to be a world center of research—he wanted it to be a work of art, and he fussed over it endlessly as the proud father he was. Perhaps it is significant in this regard that he never married.

Everything Bhabha did, he took seriously. He was an artist of some talent: a collection of his paintings and sketches has been published. His interest in architecture was deeply informed, and he would discourse at length on the architectural principles expressed at Versailles, on the facade of a famous mosque, or on the Persian concept of the apotheosis of water that lies at the heart of Moghul garden design. He published an article on da Vinci. He played the violin. He raised rare plants. Dinner parties at his penthouse apartment were elegant affairs of good food and fine wines. The guest list was carefully crafted; the conversation, always scintillating, would range from a fascinating new artist he had just discovered to government plans for the construction of a giant hydroelectric plant; from the theories of Einstein to those of Heidegger.

All his life Bhabha was fascinated by energy—both in his research and in his public policy work. It is indeed a remarkable concept. At the time of the scientific revolution, what we now call energy was a vague and fuzzy constellation of ideas. Often confused by early scientists with force or momentum, the term carried in those days little more than a sense of "oomph." Newton's *Principia* pays little attention to it—he formulated his theory not in terms of energy, but of force. Not until the nineteenth century and the development of thermodynamics did energy come into its own, achieving the status of an incorporeal, unseen and yet real attribute of the physical world.

The more time has passed, the more the concept has broadened far beyond the original intentions of its creators. Nineteenth-century physics recognized three forms: the potential energy of a soaring hawk, the kinetic energy the hawk attains by folding its wings and diving, and the chemical energy of the food it seeks. Twentieth-century physics has added more forms. We now speak of the energy of light; of massless, chargeless neutrinos; and even of empty space. But throughout all its transformations and extensions, the term has retained the essential connotation of life, of vigor. Blake said it best: "energy is eternal delight."

In his 1945 letter to the Tata family's charitable trusts, proposing creation of his research institute, Bhabha had brought up an even newer form: that of the atomic nucleus. "When nuclear energy has been successfully applied for power production in say a couple of decades from now, India will not have to look abroad for its experts but will find them ready at hand," he wrote. It was a prophetic statement, coming prior to the bombing of Hiroshima and the first public announcement of the existence of atomic power. And I think it is significant that in his proposal Bhabha explicitly linked India's need for pure research with its need for energy.

Throughout his career, he regarded the question of energy as central to the problem of national development. "No power is as expensive as no power" was how he liked to put it. In the mid-1950s he instituted a program of development of nuclear power in India. At the 1955 United Nations Conference on the Peaceful Uses of Atomic Energy, of which he was president, he gave his reasons.

Bhabha's address before this conference is a remarkable document—remarkable both for the breadth of his vision and for the conclusions he reached. He begins by dividing the range of history into three epochs. Rather than distinguishing these epochs by means a historian might use, Bhabha thought about them in a new way. He concentrated on how each period obtained its energy.

The first, preindustrial epoch drew its energy solely from muscle power. He emphasized that

> . . . a man in the course of heavy physical labor in an eight hour day can hardly turn out more than half a kilowatt-hour of useful work . . . this is to be compared with the rough figure of twenty kilowatt-hours or more of energy per person which is daily utilized in the industrially advanced countries today. It followed that a high level of physical comfort and culture could only be enjoyed by a small fraction of the population by making use of the collected surplus labor of the rest. It is sometimes forgotten that all the ancient civilizations were carried on the muscle power of slaves or of a particular class.

The second epoch was the industrial, which allowed a greatly increased level of power consumption. But contrary to most thinkers, Bhabha argued that the benefits of the industrial revolution *could not possibly be extended to the world as a whole.* There simply was not enough energy to go around—not enough energy of the normal sort, at least. So long as humanity relied on conventional technology, the developing nations would be forever prevented from attaining the level of development enjoyed in the West.

For example, Bhabha pointed out, India's per capita energy use amounted to a small fraction—the figure is one-sixtieth today—of that of the United States. Were India to increase its level of consumption to match ours, the country would utterly exhaust its coal reserves within a mere ten years! And Bhabha's point was not only confined to India. If the developing nations were to reach the level of energy use enjoyed by the developed ones, the required increase in world power generation could not possibly be provided by ordinary means. Nuclear power was the only way out of this inexorable bind. "The acquisition by man of the knowledge of how to release and use atomic energy must be recognized as the third great epoch in human history."

I found it a sad and ironic experience to read over the proceedings of this conference. People were so optimistic that nuclear energy, having devastated two cities, might now become the godsend that would resolve at a stroke humanity's deepest problems. Everything seemed possible in those days: a dramatically increased standard of living for all, the banishment of hunger, a cure for cancer. So august a periodical as *The Old Farmer's Almanac,* ever the barometer of popular opinion, referred to 1951 as Atomic Year Six.

At Trombay, a 1,200-acre plot of land across the harbor from Bombay, Bhabha presided over the construction of Asia's first nuclear reactor. It was an extraordinary effort, requiring the development of much technology new to the country—and all the more so in that, with the sole exception of the purchase of fuel elements from the United States, it was built entirely indigenously. While most European nations were purchasing their reactors from the United States, India built its own.

Throughout his career, the technological expertise for which Bhabha pressed was a purely indigenous one. He used to say that if an item of equipment was imported from abroad, all you got was that particular instrument. But if you built it yourself, an all-important lesson in expertise was learned as well. In the long run, this policy contributed immeasurably to the creation of a technical base in India. But it drastically slowed things down.

The construction of the Trombay reactor, and indeed the whole of the development of science in India, is a triumph of extraordinary magnitude. Steps that would have been trivial in the West became time-consuming excursions into research and development. The most humdrum of instruments, things we would buy off the shelf at Radio Shack, had to be built from scratch. But this required a whole network of supporting technology, a network that was mostly absent. A Western scientist who needs a minor bit of equipment can order it from a technician and get it within days. Not in India in those days, however. In the early years of the Tata Institute, Bhabha was forced to locate a promising glassblower and ship him all the way to England for training. The difficulty was that this man, lured by the superior facilities of the West, might very well never have come home again. In this instance he did; but a newspaper item during my stay in Bombay mentioned that every single graduate of the Indian Institute of Technology that year had chosen to emigrate.

India is a nation of chaos, and of numbing, stifling bureaucracy. An item of foreign equipment imported from abroad, such as the Trombay fuel elements, can be held up for months at airport customs. The scientist who shows up at the airport, seeking to pry his stuff loose, will find himself caught up in a scene of unimaginable confusion. Hordes of people mill about aimlessly: others squat motionlessly in the middle of the room, gazing remotely into the distance and seemingly resigned to spending days there. Long lines queue up before sullen, uncommunicative officials: taking his place in one, the scientist will spend hours as it inches forward, only to learn upon reaching its head that he had chosen the wrong line, or that his paperwork is not in order. Paint peels from the walls: illumination gleams weakly down from the occasional naked neon tube.

The engineers who built the reactor at Trombay worked at times around the clock. No late-night transportation was available to take them home, nor was food available at the construction site. Bhabha wished to provide these engineers with a car and food from a nearby restaurant. But both were forbidden by government regulations. He had to go all the way to the prime minister on this one.

So simple a matter as getting from here to there can be a daunting undertaking in India. Airplanes and trains are often booked to capacity for weeks in advance. As for travel by road, it is an adventure. Buses are jammed with hordes of people, oozing out of windows and perched upon the roofs. They share the road with cars, trucks, motor-scooters, bicyclists, bullock-carts, camel-carts, and pedestrians (the sidewalks are pretty much impassable). Here three men shove along a cart piled high with goods; there a bullock lies motionlessly in the very middle of the road. A monkey darts across the street: drivers lean upon their horns.

Recently an Indian academic of my acquaintance spent a sabbatical leave in America. A friend took her driving on some errands. At one point the two found themselves whizzing smoothly along on a superhighway. After a period of uneventful silence, the Indian visitor turned and addressed her friend. "Tell me," she asked, "when you drive in the States, what do you *think* about?"

There is not the slightest doubt in my mind that Bhabha would never have achieved what he achieved had it not been for his aristocratic background and personal connections. Coming as it did from a member of the family, his letter to the Tata Trust proposing his Institute of Fundamental Research must at the very least have received special attention. In leapfrogging over the bureaucracy, who he knew mattered far more than what he knew. When a group of his people building a radio telescope ran into difficulties in acquiring a site, a personal call to a state minister of industries cleared up the problem. And when Bhabha got wind one day of a plan to build a naval dockyard in his institute's front yard, ruining the view of the harbor, he used his personal connections to scuttle the project.

His most important personal connection was Nehru himself. The two had met as young men and personally liked one another. Both were of aristocratic family backgrounds, had grown up in circumstances far removed from that of the average Indian, and had spent long periods of time in the West. In later life they made every effort to meet regularly.

Bhabha's correspondence with Nehru is courteous, formal, affectionate. He always addressed the prime minister in writing as "Bhai," a term that might be translated "elder brother."

My Dear Bhai,

I returned to India yesterday after a 12-day halt in London for the Tercentenary Celebrations of the Royal Society and a 2-day halt in Paris. I attended two [lectures] and found them most stimulating. It is remarkable how by dint of immense hard work by many outstanding people one has now been able largely to reconstruct the chemical structure of the biological molecules of which living substances are made. . . .

My Dear Bhai,

I returned from Europe on Tuesday the 23rd June after a brief but very full trip.

I enjoyed the two days I spent in Cambridge in connection with the conferrment of an honorary degree on me. . . . I stayed in Cambridge at the Master's Lodge in Trinity as the guest of Lord Adrian. This was evidently a particularly good year for roses. I have never seen such a profusion of beautiful roses, as was to be found in his garden at the back adjoining the river. The two days in Cambridge, although very hectic, were most refreshing. . . .

and from Nehru:

My Dear Homi,

So you are back after collecting more honors! I liked your speech at the Cambridge Luncheon. I am glad you are staying at Bangalore to do some quiet work. . . .

Homi Bhabha the aristocrat, the "distant alp" as he was referred to by his colleague, towered over the landscape. But a mountain is not part of the normal world. He was in some ways quite unaware of India's terrible problems. I know of no more telling illustration of this fact than his proposed solution to India's population explosion. In 1959, at an address before an international Planned Parenthood conference, Bhabha called for a program of research aimed at developing a substance that when mixed with rice, would have the effect of reducing fertility of every woman in the nation by 30 percent.

Bhabha was fully aware his proposal might arouse opposition, but I think it is fair to say that he had little respect for that opposition. He understood perfectly well that people would detest such a program, but he regarded these feelings as not particularly important—they were an unfortunate stumbling block, standing in the way of his program's implementation, rather than a basic flaw in its conception. And it never seemed to have crossed his mind that children are the only form of Social Security that exists in rural India, that sons who will care for them in old age are all that stands between aging villagers and starvation. It is characteristic that his approach to the problem of population—of any social problem—took little account of social realities.

Rather, he thought in terms of research. Bhabha had an unbounded faith in the power of science as an instrument of social change. So far as he was concerned, the problem of national development was nothing more than a problem of the development of modern technology, and he did everything in his power to foster that development. At the time of his death, he was working on a survey of India's electronics industry, ranging from the most inexpensive radio to the most sophisticated silicon chip. He formulated policies to protect its vast deposits of thorium, used in nuclear reactors and medical technology. He inaugurated the Indian space program.

Control of science in India is highly centralized, and its directors are largely answerable only to themselves. Toward the end of his life, Bhabha accumulated enormous powers. He was director of the Tata Institute of Fundamental Research, director of the nuclear reactor complex at Trombay (renamed the Bhabha Atomic Research Centre after his death), secretary to the government of India in the Department of Atomic Energy and as such also ex-officio chairman of India's Atomic Energy Commission, and chairman of the Scientific Advisory Committee to the Cabinet. In the year before his death, the budget under his direct control amounted to 115 million rupees—more than $6 million at today's rate of exchange. On the international scene he was president of the first United Nations conference on the Peaceful Uses of Atomic Energy, chairman of the Union of Pure and Applied Physics, and of the International Atomic Energy Agency.

In thinking of this man, the image perpetually rises to my mind of one of those great, larger-than-life figures of the Renaissance: laying out plans for a new city one day, waging war on the next, writing a sonnet on the third. When the International Atomic Energy Agency was created and he was named its director, Bhabha established the headquarters in Vienna at least in part because of that city's cultural life. A visit to London for a meeting of the Royal Society would be combined with a meeting with a government minister; in Chicago for a conference with the director of a giant hydroelectric plant, he might drop in on a well-known art collector.

These three quotations from Homi Bhabha's writings illustrate his range:

> Art, music, poetry, and everything else that I do have this one purpose—increasing the intensity of my consciousness and life.

> [The Canadian-Indian nuclear reactor] project will have to be handled by the Department of Atomic Energy, not

only at the technical level but at the inter-governmental level, and the Agreement will have to be signed on behalf of India as the Secretary of this Department . . . unless this action is taken, all inter-governmental correspondence will have to be routed through the Department of Economic Affairs.

The permanent things in nature are certain generalized concepts like energy, momentum, angular momentum and electric charge, which are always conserved, while the actual elementary particles themselves are but the transitory embodiments of their metamorphoses.

Bhabha died in 1966 in an airplane crash on a trip to Europe. The news reached India just as Indira Gandhi was being sworn in as prime minister. A postage stamp was issued in his honor. The first day cover shows him brooding mildly over Trombay, his nuclear city: beside him is an artist's palette, and beneath, the "Ode to Joy" theme from Beethoven's Ninth Symphony.

5

Luie's Gadgets

Luis Alvarez and the Extinction
of the Dinosaurs

Several years after the close of the Second World War, the physicist Luis Alvarez realized that he was having trouble at lunchtimes. In the cafeteria at the University of California at Berkeley a subtle but all-too-common segregation of personnel had developed. Over in one corner, the young people sat by themselves, talking physics. In the other corner sat the older and wiser heads. They were talking about the good old days. Alvarez recognized with a shock that he had been unconsciously avoiding sitting with the young, up-and-coming generation. From being a hotshot himself, he had imperceptibly slipped into being a has-been. He was forty years old.

Not an uncommon situation. But Alvarez's response was unique. He hired two young graduate students, and he told them that although on paper he would be their instructor, in reality *they* would be *his* instructors. He pushed his desk close to theirs and placed himself entirely in their hands. The two students looked at each other and swallowed hard—and then they set to work, teaching their boss the new physics.

Such vigorous, crisp moves were part and parcel of Alvarez's style. Nor did the role reversal appear to particularly bother him: there was not an ounce of pretension to the man, nor an ounce of sentimentality. He had been born in 1911 in San Francisco and he died in 1988. His father was a doctor and a nationally known syndicated columnist for a chain of newspapers, writing regularly on health and medical matters. Luis's paternal grandfather was Spanish, but otherwise there was nothing Hispanic about the family.

His greatest forte was an ability to push the state of the art of experimental physics to its very limits. He achieved this by combining an intense inventiveness with a deep familiarity with the latest technological advances. Alvarez was, in fact, in love with technology, and he was one of its most skilled practitioners. Commenting at one point on the Mad Hatter of Lewis Carroll's *Alice's Adventures in Wonderland,* he noted that fumes from the liquid mercury used in forming felt often caused brain damage. This little tidbit was an element of Carroll's creation that Alvarez found worth mentioning. He was the quintessential gadgeteer.

He could no more refrain from invention than he could from breathing. In the midst of an illness so severe as to incapacitate him for most of a summer, he amused himself by trying to build a better detector of gallstones. In the midst of an exhausting round of meetings in Washington while lobbying for development of the hydrogen bomb, he found time to stop off at RCA to see a demonstration of the newly invented color television— within months he was consultant to a firm founded to develop a commercially marketable set. By the time of his death he held more than forty patents.

During the Second World War, Alvarez worked for several years at MIT on the development of radar. So prolific were his powers of invention that he was made head of a group whose sole function was to bring his new ideas into fruition. The group's official designation was Division 7, but everyone knew it as Luie's Gadgets. There he invented the Ground Controlled Approach radar system for guiding airplanes to landing in bad weather. He invented a radar antiaircraft system. He invented the radar bombsight. He invented VIXEN, a radar set for fighter airplanes designed to find enemy submarines: it is one of the prettiest gadgets I know.

Radar works by emitting a sharp pulse of radio waves and detecting the echo as it bounces off the target. It is like locating a nearby cliff in the dark by clapping your hands. The problem is that the cliff can "detect" the sound of the clap more easily than you can detect the sound of the echo. In the early years of the war, allied pilots found their primitive radar sets effectively useless in waging war against enemy submarines. They would detect the submarines easily enough, but the emitted radar signals also acted as a beacon, advertising the fighter's presence to the enemy. As the plane neared its target, the beacon would grow yet stronger, and the U-boat would promptly dive.

Alvarez could think of no way to mask this beacon. But he did think of a way to mask its significance. He guessed that the real problem was not so much the presence of the radar pulses. The real problem was *the fact that they grew stronger as the fighter approached.* So he designed a radar set to emit a series of pulses of steadily *decreasing* strength. In his design, he was helped by the technical peculiarity that the strength of the beacon alerting the submarine to the presence of the aircraft depended on the inverse square of its distance, but that of the return echo as the inverse fourth power. This made it possible to adjust things so that the aircraft would receive an ever more powerful echo as it approached its target, even though at the same time the warning beacon received by the submarine was growing steadily weaker. Lulled into a false sense of security by this apparent evidence that the aircraft was flying away from him, the U-boat commander would not give the order to dive until suddenly pounced upon by the fighter.

Alvarez spent the latter part of the war at Los Alamos, working on the atomic bomb. He took part in the mission that dropped the Hiroshima weapon. In his autobiography, *Alvarez,* he describes the event:

> The bomb took forty-five seconds to drop thirty thousand feet to its detonation point, our three parachute gauges drifting down above. For half that time we were diving away in a two-g turn. . . . Suddenly a bright flash lit the compartment. . . . A few moments later two sharp shocks slammed the plane.

After we secured our equipment, we left our cramped quarters and looked out the window for the first time over Japan. By then [the pilot] was heading back toward Hiroshima, and the top of the mushroom cloud had reached our altitude. I looked in vain for the city that had been our target. The cloud seemed to be rising out of a wooded area devoid of population. My friend and teacher Ernest Lawrence had expended great energy and hundreds of millions of dollars building the machines that separated the U-235 for the Little Boy bomb. I thought the bombardier had missed the city by miles—had dumped Ernest's precious bomb out in the empty countryside—and I wondered how we would ever explain such a failure to him. [The pilot] shortly dispelled my doubts. The aiming had been excellent, he reported: Hiroshima was destroyed.

We flew once around the mushroom cloud and then headed for Tinian. Japan looked peaceful from seven miles up.

There is only one emotion expressed in this passage: a momentary horror that one of the two atomic bombs in existence might have been wasted. Conspicuously missing is any expression of deep remorse for having taken part in the destruction of Hiroshima. As a matter of fact, Alvarez was proud of his role in the development of the atom bomb. One of the photographs he chose to include in his autobiography was of him holding a model of a B-29 as his son sat on his lap; and on several occasions, he expressed scorn for those colleagues at Los Alamos who did express remorse.

One might be tempted to interpret all this as yet another instance of the quintessential amoral scientist—hard at work in the laboratory, totally engrossed with his gadgets and far too concerned with them to worry over their ethical consequences. But I would resist this view. Alvarez thought long and hard about the morality of the atomic bombs, and he concluded that their use had been wholly ethical. On several occasions he referred to the massive loss of life, both American and Japanese, that their use averted; and he pointed to the dreadful threat of a third world war, which appears, for the time at least, to have been held at bay by the thermonuclear threat.

But the central fact I would emphasize about his years at Los Alamos is that Alvarez enjoyed them. He liked working on the bomb. He liked the challenge of clearing the ever more daunting hurdles, of pushing forward into new technologies.

The career transition after the war in which his two students played so great a part was into the field of elementary particle physics. He was working at an accelerator that sped protons to nearly the velocity of light and slammed them into a target. Out of these collisions would emerge a spray of new elementary particles. Incomparably smaller than atoms, however, they were hard to study. Alvarez plunged into the analysis of new means of detecting them.

In a corridor of the Berkeley physics building in those days stood a box: it was several feet across, and covered with a sheet of glass. Within, silvery clouds of alcohol droplets drifted, suspended in a gas. Every few seconds, suddenly and without warning, a track would appear in these clouds—a straight line of droplets. Each track was triggered by the passage of a cosmic ray, Bhabha's elementary particles from space. An impressive demonstration: morning, noon, and night the chamber continued developing tracks, documenting the bombardment of the rays.

The box was called a cloud chamber. Within it, the alcohol had been dissolved in air, and in such quantities as to be continually condensing into tiny droplets. But the process of condensation was facilitated by the passage of the elementary particles: whenever a cosmic ray passed through, droplets would quickly appear. Thus the tracks. The chamber was functioning as a detector of elementary particles.

Other detectors used in those days employed stacks of photographic plates, the grains of which would be exposed by passage of the particle. But all suffered from a variety of problems. Analysis of the tracks left on the plates was tedious, inefficient, and required the use of a microscope. After a mere few hours spent analyzing their data, physicists would wander around dazed, rubbing their eyes. As for the cloud chamber, its sensitive region was only a few inches deep—not enough to get a good look at the particles as they whizzed through. There were other chambers that had a larger sensitive volume, but they suffered from the defect of uncomfortably long dead times: minutes were often required after one detection for the alcohol to redissolve. Accelerators, on the other hand, emitted a spray of particles more like once a second. These inefficiencies of detection were one of the primary bottlenecks impeding the study of elementary particles.

One day at a conference, Alvarez found himself seated at a table with a number of cronies from Los Alamos. Predictably, the conversation revolved around the war. But seated next to him was a young postdoctoral fellow named Don Glaser. Alvarez turned to Glaser and began talking physics.

Glaser told him about a new particle detector he had invented—not a cloud chamber, but a bubble chamber. It consisted of a liquid continually heated to greater-than-boiling temperature, but which was prevented from boiling by high pressure. Upon expansion of a piston, the pressure would drop and the liquid would boil. And as with the process of condensation, so too with vaporization: elementary particles would help along the boiling, and along their tracks there appeared a string of tiny bubbles.

Alvarez realized immediately that bubble chambers possessed enormous advantages compared to cloud chambers. At a stroke, many of the difficulties plaguing the old design could be evaded by Glaser's gadget. Glaser had employed ether as his sensitive liquid; but in his hotel room that night, Alvarez worked out with colleagues plans for a chamber based on liquid hydrogen. Such a chamber could be built with far larger sensitive volumes, it could be recycled not once a minute but once a second, and because the liquid hydrogen was of such a high density, it could function as the target against which new elementary particles would be created.

Alvarez's first bubble chambers were a matter of inches in size. But ultimately, he and his group constructed a monster fully 6 feet long. The development of such chambers was no

easy task. It required the invention of not one but a whole range
of new technologies. In later life he commented that his group
was not pushing the state of the art so much as sailing over the
edge—a daunting business, and the stuff of sleepless nights.
"Dashing off into the scientific unknown isn't for the faint-
hearted," he wrote in his autobiography. "You must believe you
can find a way." At one point he and his people knowingly designed themselves into a corner, trusting they would eventually
come up with a way around a seemingly insurmountable obstacle plaguing their design. But all this was just the sort of daring,
risky venture Alvarez liked best. It appealed to his can-do nature.

Cleanliness was an important issue. Bubbles formed not just
around elementary particles, but about the most microscopic of
imperfections. A speck of dust or a bump on a wall could trigger
the boiling of the liquid, jamming the detector with spurious
signals. So chambers had to be clean, utterly clean. Their interior surfaces needed to be mirror-smooth, and the difficulty of
achieving this limited their sizes to matters of inches. But Alvarez found a simple and characteristically pretty way around
this restriction. While he kept their windows spotless, the remaining surfaces of his chambers he allowed to be dirty. The
spurious bubbles then formed in profusion—but precisely because his chambers were so gigantic, the bubbles were off to one
side of the windows, and safely out of sight. Everyone knew in
those days that you could never make a 6-foot chamber clean.
What Alvarez realized was that you did not have to make a 6-
foot chamber clean.

The detectors he and his group built transformed the field.
They produced a flood of data on elementary particles. But ironically, their very success posed a problem. As anyone who has
seen a reproduction can testify, even a single bubble chamber
photograph contains a wealth of information—tracks upon
tracks, each one curving this way or that, and all needing careful analysis. Furthermore, his chambers could churn out hundreds of thousands of such photographs in a day. No one would
be capable of dealing with such an inundation. Alvarez realized
that computers were the only means of handling all the data.
He and his group developed automated methods of scanning

the photographs. The computer programs they wrote were precursors to the modern fields of artificial intelligence and pattern recognition by machine.

The liquid hydrogen their chambers employed can be deadly dangerous. If even the slightest leak develops and it gets out into the air, the stuff is ferociously explosive. The Cambridge Electron Accelerator, just off Harvard Square and operated by a Harvard-MIT collaboration, once suffered a catastrophic explosion of its bubble chamber. The roof was blown upward and then plunged down upon the support pillars; the pillars collapsed under the impact, and thousands of tons of concrete rained down upon the accelerator. All the cables caught fire and poisonous smoke spread throughout the laboratory. But not once in all their years of bubble chamber work did the Alvarez group suffer a serious accident.

The 1950s were the glory days of elementary particle physics at Berkeley. With their new detectors, Alvarez and his people dominated the field. There were periods in which only they had access to certain phenomena. The theoreticians hovered about, peering excitedly over their shoulders as the data streamed in. Particle after particle was discovered, particles whose properties were not just new and unfamiliar, but for which whole new categories of explanation were required.

Physicists were elucidating the ultimate building blocks of matter. It is worth pausing to marvel at the fact that such fundamental units of construction even exist. That objects as diverse as stones and trees, automobiles, our bodies, and the very stars themselves—that all these are built from the same elementary particles must rank among the most remarkable of discoveries.

But the question of what these fundamental units are has undergone a considerable evolution. The Greeks argued that the world was made of earth, air, fire, and water. More recently, people felt that atoms were the ultimate building blocks—that is what Boltzmann had fought for. But the multitude of chemical elements, each with its own atom, posed a problem for the atomic hypothesis, at least if you took the view that at heart, nature was simple. Furthermore, by the early years of this century it was recognized that atoms themselves were made of yet smaller particles. These were the protons, neutrons, and electrons; and I was taught in college that to these three particles—we called them "elementary" back then—all the bewildering complexity of the world could be reduced. A few other particles were known: the pion, the muon. When I thought about these interlopers, I was forced to admit that I had not the slightest idea what function they played in the nature of matter . . . but to tell the truth, I did not think about them very much. The picture was too pretty.

The 1950s and 1960s changed all that. The few particles known when I was a student turned out to be the mere tip of an iceberg. New particle after new particle was discovered, and the roster kept growing. Furthermore, as it did, patterns began to emerge, patterns whose existence had previously been unsuspected. Gradually it came to be realized that the very simplicity of the early view had been an obstacle to our understanding.

The discoveries made possible by the bubble chamber led to a revolution in ideas of the ultimate nature of matter. Protons and neutrons are now known to be composed of yet smaller particles called quarks. The success of the quark model in accounting for subatomic phenomena is impressive. But it is worth noting that this picture, too, suffers from certain difficulties. The number of particles it takes to be truly elementary, for example, has grown considerably over the years from an initial three to a present twelve. Newer and more speculative ideas increase the number of elementary particles still further. Is yet a further proliferation in store?

If history is any guide, future work may well reveal yet a deeper level of the organization of matter. Such particles as quarks, which we now think of as elementary, may themselves turn out to be made of something else, something even smaller. But will this continual redefinition of our ideas ever end? *Are* there any fundamental building blocks of matter? Or is the process of peeling away successive layers of the onion never ending? And another question: if we finally do discover the truly elementary particles, how many of them will there be? Just a few—or hundreds? Surely, it is only our prejudice that at heart, the world is simple. But perhaps in the long run it will turn out to be complex.

In the spring of 1954 Alvarez testified at the hearings in which J. Robert Oppenheimer, director of the research effort at Los Alamos, was stripped of his security clearance.

One of the first questions Alvarez was asked at the hearings was whether he wanted to testify. He replied "I certainly find it an unpleasant duty, but I consider it to be a duty to be here." And indeed, the sense of duty, and of conflicting duties, marks much of his attitude throughout this episode. He had not been forced to testify—not subpoenaed. Rather, the committee had requested him to appear, and he was perfectly free to refuse. His job at Berkeley would not have been jeopardized had he stayed away: he had never, so far as I can tell, so much as flirted with radical politics. Ernest O. Lawrence, simultaneously his friend, mentor, and boss, and a man whom Alvarez deeply admired, had refused to testify and had told Alvarez not to. In his autobiography Alvarez speaks of "disobeying" Lawrence in this matter—a term I find somewhat curious, more appropriate to a youngster than to someone of Alvarez's stature at the time. But Lewis Strauss, Chairman of the Atomic Energy Commission, had phoned him at his home and appealed to his sense of duty, predicting that he wouldn't be able to look himself in the mirror for the rest of his life if he refused. Strauss seems to have known his man, and Alvarez agreed.

His testimony was relatively short. Much of it concerned his efforts to push for development of the hydrogen bomb and his disagreements with Oppenheimer, who opposed it. He expressed his admiration for Oppenheimer—and for Oppenheimer's opponent Edward Teller as well. He described Oppenheimer as "one of the most persuasive men that has ever lived," and in the core of his testimony he spoke of his sense that Oppenheimer had exerted an all-pervasive influence in creating opposition among other scientists to the hydrogen bomb:

Every time I have found a person who [opposed the bomb], I have seen Dr. Oppenheimer's influence on that person's mind. I don't think there is anything wrong with this. I would certainly try to persuade people of my point of view. . . .

DR. EVANS: It doesn't mean he was disloyal?
ALVAREZ: Absolutely not, sir.

This marks the only point at which Alvarez was asked this crucial question.

At one juncture the tone of these sordid hearings is relieved by a dim ray of humor. The discussion has been centering on Alvarez's view of the morality of the bomb:

DR. EVANS: Don't we always have moral scruples when a new weapon is produced?
ALVAREZ: That is a question I can't answer, sir.
DR. EVANS: After the battle of Hastings, a little before my time—
MR. SILVERMAN: Would you give the time, sir?
DR. EVANS: I cannot give the time, but it was before I was born.
MR. SILVERMAN: That is 1066, sir.

The Oppenheimer hearings were one of the most disgraceful episodes of their time. But does this mean that Alvarez should not have testified, or that he should not have said what he did? His supporters have always pointed to the fact that in his testimony he defended Oppenheimer's loyalty. As one too young to have been politically conscious during the episode, I do not feel that my comments carry much weight. But in talking to older colleagues, people who lived through those times, I have heard every shade of opinion expressed—and all of it expressed vehemently. One colleague who knew him well has told me that Alvarez should have refused to testify: another, who knew him equally well, that it took more courage to testify than it would have to have refused. Yet another colleague described the hearings as a witch hunt, and he said that anyone who did not support Oppenheimer at every juncture was open to being used as a tool against him—and that Alvarez knew this perfectly well.

The Oppenheimer hearings polarized physicists as no other event in recent history. In a famous episode, Oppenheimer's opponent, Edward Teller, encountered two friends while walking with a colleague shortly after the hearings: the friends shook the colleague's hand, but they refused even to acknowledge Teller's existence. This bitterness, this rancor, has continued unabated to our day—and because of his testimony, Alvarez is generally regarded as having belonging to the Teller camp. "Luis Alvarez: wasn't he that son of a bitch out at Berkeley?" inquired a colleague when I mentioned his name not long ago.

"Was he a son of a bitch?" I responded.

"Well, I never knew him personally, but I understand that he was."

In his autobiography Alvarez wrote that he was not aware of having lost any friends as a result of his testimony. Nor do I believe that the prospect would have stopped him from testifying. He was, in fact, a person of immense self-confidence. Once he made up his mind about an issue, he was not one to stew over it.

Alvarez was crisp. Alvarez was cool. He was gruff and plain-spoken: devoid of pretension, he had no use for it in others. His wit was sharp and at times cruel, his intelligence utterly clear and to the point. There was something almost military about him. Indeed, in contrast to most scientists, Alvarez admired and respected the military. At Los Alamos he enjoyed singing soldiers' songs, and during a wartime stay in England, he felt honored to associate with the pilots fighting the Battle of Britain. In the early 1960s he chaired a committee on limited war, and he invented a night-sight scope for use in Vietnam.

His hair was intensely blond, his features elegant and finely chiselled. He was a man of strong opinions: his political views were personal, idiosyncratic, and characteristic of no particular line. But one of his opinions was that only the opinions of an expert were worth listening to. Awakened by the telephone in the middle of one night in 1968, he was told by a reporter that he had just won the Nobel Prize for his bubble chamber work. Lying in bed after hanging up, he talked things over with his wife. Then and there he resolved never to sign petitions or otherwise express a political position solely on the basis of the prize. Expertise as a physicist gave him no right to make pronouncements about anything other than physics, he felt.

He could be kind and charming. He would go any lengths to help someone he respected, and he was the sort of person who would enthusiastically shake the hand of someone who had had a better idea than his own. When he went to Stockholm for the Nobel ceremony, he acknowledged his indebtedness by inviting members of his research group to accompany him and personally paying their travel expenses.

But he did not always make a point of setting others at their ease, and there were those who found him intimidating. He was a man who formed strong judgments of people, and those whom he considered unworthy were utterly dismissed from consideration. Nor was he one to hide his opinions—once he summed up the position of a colleague who disagreed with him by saying, "Bill's theory is that our theory is wrong."

Alvarez's autobiography is written in a crisp, no-nonsense style that I find utterly refreshing. There is much wisdom in it, and it is presented with a complete absence of pretension and sentimentality. Narrowly having missed discovering atomic fission, he comments, "I'm probably lucky to have missed the discovery of fission. I doubt I had the maturity at twenty-seven to handle the burden of having made one of modern science's greatest discoveries." Elsewhere he recounts how he independently rediscovered artificial radioactivity. His only comment was "If I hadn't been three years too late, I would have certainly told someone about it."

Alvarez's bubble chamber work marked an important point in the evolution of his research style—and in the style of physics as a whole. It required a large team, in which the individual scientist became something more like a manager than an independent agent. By now this evolution has proceeded so far that it has totally transformed the nature of research: chapter 7 is devoted to a man whose career has been dramatically affected by this change. As much as anyone else, it was Alvarez who was instrumental in leading physics into the era of giant consortiums. But he himself was in many ways most comfortable working alone, or in small groups.

In later years his groups grew smaller, the projects more idio-syncratic, inventive, and to me at least, more delightful. Per-haps the most delightful of all involved, sadly, a ghastly event: the assassination of President Kennedy. Alvarez first became aware of the famous Zapruder film of the assassination when in-dividual frames were published by *Life* magazine. He spent the better part of that Thanksgiving weekend poring over the issue, studying each individual picture as he had once studied bubble chamber photographs.

His attention was drawn by a set of brilliant, parallel streaks marking certain of the frames. They were images of the Sun, re-flected off the body of the presidential limousine. Most of the time Zapruder had tracked the motorcade perfectly with his camera, and the reflections appeared in the movie as tiny points of light. But in a few of the frames, the points were elongated into streaks.

It occurred to Alvarez that these elongations would have arisen had Zapruder involuntarily jerked at these moments—and that he would have jerked had he been startled by gun-shots. His suspicion was confirmed when a news organization staged a reenactment of the event and discovered similar streaks in their films. Thus, Alvarez realized, he could count the num-ber of shots fired at Kennedy by counting the number of frames showing streaks. His results discredited conspiracy theories of the assassination. He was also able to show that a sudden back-ward snap of the president's head arose not because he had been hit by two bullets from opposite directions, but from the momentum carried by the jet of brain matter expelled by the bullet, again arguing against the conspiracy theory.

During the early 1970s, Alvarez spent a good deal of time on the phone talking with his son Walter. Walter was a geologist, and they were doing what scientists like best: bouncing ideas off one another, trying out schemes for this and that. Out of their dis-cussions there developed the last major shift in Luis's career, and a remarkable new insight into the cause of the extinction of the dinosaurs. Like many great discoveries, this one came about through a chain of hard work, stray circumstances, and lucky breaks.

The dinosaurs, largest land animals ever to have existed, were an extraordinarily successful group that survived for 120 million years before suddenly going extinct. (For comparison, humanity has existed for something like 1 percent as long.) For decades their extinction, which occurred about 65 million years ago, constituted one of the most famous mysteries in all of science. There was a proposal that their eggs were eaten by the mammals. There was a proposal that they died of diseases to which they had not developed resistance, contracted when they migrated across newly formed land bridges. There was a proposal that they died of the cold, and another one that they died of too much heat. Someone once said that they died of constipation.

The geological stratum marking the extinction event is known as the K/T (for Cretaceous/Tertiary) boundary. Walter's interest in the boundary was "an offshoot of an offshoot of an offshoot," as he told me. In 1973 he was interested not in dinosaurs but in a completely different question, one that took him to a gorge outside Gubbio, a small town north of Rome. The data he obtained from the Gubbio gorge turned out to be of little use for his original project. But he realized that the rocks there contained a wealth of other information. He returned to Gubbio several summers for other purposes.

With him was a paleontologist who had done her thesis on the K/T boundary. Often the two would walk up and down the gorge chatting, and every time they passed the K/T boundary, she would tell him something about it. It was, indeed, a striking geological formation. The rocks above and below were a shade of pink. But for about a foot below the boundary, they were bleached white, and the boundary itself was marked by a centimeter-thick layer of clay. That clay layer was the residuum of the epoch in which something occurred that wiped out the dinosaurs.

In their telephone conversations, Walter mentioned to his father that an important question was how great a span of time this clay represented. How long had the epoch of extinction lasted? A million years? A week? Eventually they hit upon an ingenious method of measuring the duration of this interval. They realized that there was a clock whose "ticks" would have been preserved within the clay layer.

These ticks involved meteors. The Alvarezes' method was to use the rate of infall of meteors as their clock. Once in a long while, relatively large objects strike the Earth. But far more often, tiny ones do. Indeed, every hour of the day microscopic meteors—tiny bits of fluff, grains of sand—impinge upon the Earth. Halted in the upper atmosphere, their remnants drift unnoticed down upon our heads in a continual rain of extraterrestrial material.

Contained within the rain are traces of the element iridium. Rare upon the surface of the Earth, iridium is more common in meteors. The Alvarezes' plan was to measure its concentration in the boundary layer. By comparing this with the known rate of infall in meteors, they would be able to determine the interval of time over which the layer had formed.

Crucial to this scheme was their ability to measure accurately the concentration of iridium in rocks. There had been a previous measurement of this concentration, and based upon it, the Alvarezes felt there was enough of it for their technique to work. It later developed that this measurement had been wrong: had the previous investigators done the job correctly, Walter and Luis would have decided their scheme had little chance of success. They would have abandoned the project, and never would have made their discovery.

Frank Asaro and Helen Michel joined them in their project. They operated a lab at Berkeley that measured the iridium concentrations. Asaro, in fact, was dubious about the scheme and limited the number of samples to be measured to twelve. Walter carefully selected his samples, turned them over—and promptly forgot about the whole thing. Time passed . . . lots of it.

Eventually, a full eight months later, Asaro walked into his office and told him that he had his results—but that Walter wasn't going to believe them. They had found that the concentration of iridium at the K/T boundary was hundreds of times higher than in the adjacent rocks. Worst of all, independent evidence existed to show that this extraordinarily high concentration could not possibly indicate a correspondingly huge period of time over which the boundary clay had formed. There was only one conclusion: something must have happened to interrupt the normally steady ticking of the meteor "clock."

Walter and Luis never did succeed in measuring the duration of the extinction event. But they had inadvertently stumbled upon something far more interesting. By now the iridium excess has been found at the K/T boundary throughout the world, both on land and under the sea. Some agency must have brought an enormous influx of this particular element to the Earth just when the dinosaurs went extinct, and it appears to spread the stuff uniformly about the globe. What could have done such a thing? For a time the Alvarezes considered the possibility that the iridium had been ejected in an explosion of a nearby star. But the pattern of abundances of various isotopes argued against such an explanation. Eventually they began thinking about the impact of a giant object.

Such an object would have had to have been close to the size of Manhattan to account for the amount of iridium in the layer. That is the size of a comet, or an asteroid. The impact of such a body would have been a cataclysm of unparalleled magnitude. Traveling at something like 100,000 miles per hour, its top would have still been in the upper reaches of the atmosphere at the moment its base struck the ground. The thing would have dug a crater as much as 100 miles in diameter, and released 100 billion times more energy than a giant earthquake. Had it struck in the ocean, it would have punched all the way down to the ocean floor—a hammer slamming down upon water spilled upon a table.

But devastating as these effects were, they were not enough to account for the worldwide nature of the extinction. Dinosaurs on, say, the other side of the globe would hardly have noticed the impact. What, then, could have been so catastrophic to the dinosaurs about the collision? This conundrum puzzled the team for some time. Asaro recalls that Luis invented a new scheme every week for six weeks, shooting down each one in its turn.

Eventually he hit upon the idea of global darkness. The impact, he calculated, would have pulverized enormous quantities of rock, and blasted it upward into the atmosphere. Spread by the force of the blast, the debris would have circled entirely about the Earth, and it would have remained within the atmosphere for months. While there, Luis argued, it would have blocked out the sunlight. A worldwide night would have ensued, a night lasting months. Night—and also cold. The cold could have directly killed off the dinosaurs. Alternatively, it would also have interrupted the growing cycle of plants, and so disrupted the food chain upon which they depended. Thus the extinction.

The impact theory of dinosaur extinctions triggered one of the most intense scientific debates of recent years. In 1979 Walter attended a conference on the K/T boundary to present their ideas. "[I] thought the paleontologists would be delighted to learn what caused the extinctions," Luis wrote in his autobiography. "Walt knew better." But even Walter was surprised at the vehemence of the opposition. "Some people at the conference were exhilarated and excited," he told me. "But most of the people working on the problem just wished we'd go away."

Resistance to new ideas is commonplace, within science just as much as without it. I myself am not immune to this resistance. I am not even so sure it is always a bad thing. The ideas of one's field, won at such a great cost of time and effort, won by untold numbers of workers—these are precious. The sense arises that they are one's own property, zealously to be guarded against outside depredations. Science, in fact, is a community, and like all communities, scientists have a tendency to close ranks against outsiders. In this case it was Luis who was the outsider, a physicist butting into the ranks of the paleontologists. And it did not help that the impact theory sounded so much like science fiction.

But I believe there is more to the matter than this. It is not enough to explain the intensity of the opposition triggered by the Alvarezes' proposal. People were amused, condescending, disgusted. It was if the Alvarezes had stepped on a raw nerve. In fact, my guess is that this is exactly what they had done. I be-

lieve that their proposal stirred up a bitter historical memory—
the memory of the painful and contentious birth of geology. It
turned back the clock, and it returned to an old and totally discredited form of explanation in geology.

The birth of geology in the late eighteenth century was
marked by a furious debate between two schools of thought.
They were known as uniformitarianism and catastrophism. The
uniformitarians held that every geological formation could be
understood as having been caused by processes underway at the
present epoch: mountains by slow uplift, canyons by incre-
mental erosion. The catastrophists, on the other hand, resisted
this notion. "No river has deepened its channel one foot,"
wrote George Greenough, founder and first president of the Geological Society of London. "No amount of time could enable
Nature to work wonders." Rather, these wonders were evidence
of ancient cataclysms.

Greenough's use of the word "wonders" here was no accident. The aim of many of the catastrophists was religious, and
they made no bones about it. These people wanted to reconcile
science with religion. They wanted to square the emerging discoveries of paleontology with the biblical account of creation.
Seashells found in the Alps were presented as evidence of
Noah's flood; fossilized remains of animals never before seen as
evidence of a time before Genesis, long since destroyed by catastrophes.

Furthermore, these catastrophes were held to lie outside the
normally operating laws of nature. They were special, unique
events. With this, the debate between catastrophism and uniformitarianism became a debate on the very possibility of science. Are the things we find about us today sufficient to explain
the world? Or are special explanations required, explanations
involving processes lying outside the normal functioning of nature? To my mind the most penetrating of all the critiques of
catastrophism was leveled by the Scottish geologist James Hut-
ton, who emphasized that the tenets of that doctrine rendered
the world fundamentally incomprehensible. "If a stone which
fell today were to rise tomorrow," he wrote, "there would be an
end of natural philosophy. Our principles would fail, and we

Annie Jump Cannon at work. Courtesy of the Harvard University Archives, from the collection of the Harvard College Observatory.

Cecilia Payne-Gaposchkin in a meditative mood. Courtesy of Katherine Haramundanis.

The Ladies of Observatory Hill. Annie Jump Cannon is second from the left in the middle row, Cecilia Payne-Gaposchkin second from the left in the rear row. Courtesy of the Harvard College Observatory.

Ludwig Boltzmann at his most severe. University of Vienna, courtesy of AIP Emilio Segre Visual Archives.

Ludwig Boltzmann, cartoon by K. Przibram. University of Vienna, courtesy of AIP Emilio Segre Visual Archives.

George Gamow in an expansive mood. Courtesy of AIP Emilio Segre Visual Archives.

Homi Bhabha. The Western-style coat and tie were fixtures. Courtesy of AIP Emilio Segre Visual Archives, Physics Today Collection.

Luis Alvarez on his last day at MIT before departing for Los Alamos. Courtesy of the Lawrence Berkeley National Laboratory.

Luis Alvarez in his office in later years. Courtesy of the Lawrence Berkeley National Laboratory.

Richard Feynman. Courtesy of the Archives, California Institute of Technology.

Richard Feynman explaining a point to one of his heroes, P. A. M. Dirac. Courtesy of the Archives, California Institute of Technology.

John Huchra with hat. Courtesy of John Huchra.

Martin Perl in his office at SLAC. Courtesy of Martin Perl.

SLAC, the Stanford Linear Accelerator Center. The accelerator runs beneath the two-mile-long building in the upper right; the particles are studied in detectors housed in the various buildings to the center and lower left. Courtesy of Martin Perl.

Margaret Geller with a copy of the first "slice of the universe." Courtesy of Margaret Geller.

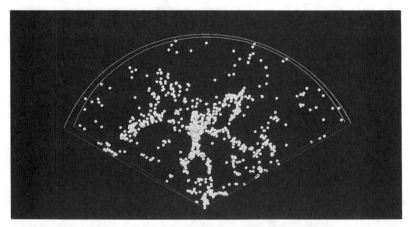

Our address in the universe. The results of Geller and Huchra's first survey of the universe. We sit at the apex of the "pie slice." Courtesy of Margaret Geller.

would be no longer investigating the rules of nature from our observations."

I do not want to give the impression that the catastrophists were fools. That would make things too easy. But the reality is that they were excellent, serious scientists: Greenough, after all, was founder and president of the Geological Society. The catastrophists had good evidence for many of their claims. Extinctions are an example, for the geological record often shows sudden discontinuities in fossil types. But Darwin always claimed that these discontinuities were illusions, induced by gaps in the geological record: not every epoch was well represented by fossils, and such abrupt shifts were to be expected. In the long run, arguments such as these triumphed, and catastrophism was demoted to the oblivion that awaits all defeated theories.

My hunch is that the impact theory of dinosaur extinctions encountered such intense resistance because it appeared on first glance to be reinvoking catastrophism. In reality, of course, it was not—their postulated impact was a catastrophe, all right, but it was not one lying outside the normal workings of nature. The Alvarezes' proposal escapes the criticisms leveled against the biblical catastrophists by widening the sphere of discussion to include other bodies of the solar system. Catastrophic impacts, far from being supernatural events, then become mere traffic accidents. Furthermore, while the geological record of extinctions is indeed often spotty, as Darwin had argued, in some instances it is not—and the K/T extinction is one of these instances. Walter Alvarez is of the opinion that the time has come for geologists to realize that a full understanding of the geological record will require both gradual and sudden processes.

To my mind, one of the most amazing aspects of the birth of the impact theory of dinosaur extinctions is that *this was not the first time it had been proposed*. Several years earlier the chemist Harold Urey had put forward essentially the same idea. But nobody had paid the slightest attention to him. By the time of the discovery of the iridium layer, it would have been hard to find a geologist who even remembered his proposal. But why? It could not be that Urey was regarded with disdain—he too was a Nobel Prize winner.

It was the iridium layer that Urey had lacked, and it was this layer that gave the theory its compelling force. This immense power of a single observation was what Luis Alvarez loved about experimental physics. Throughout his career, this is what drove him: the marvelous power of facts and of quantitative measurements, and above all else the beauty and cleanliness of hard data.

Early in his career, much of his work in nuclear physics had been motivated by one particular research article written by a noted authority in the field. Whenever the author had said a phenomenon would never be observed, Alvarez set out to prove him wrong. This, he said, would have made both of them happy.

6

All Genius and All Buffoon

Richard Feynman

Listen—I've got a question for you. What is the energy of nothing at all?

It's a silly question, of course. You don't have to be a scientist to know that energy is always the energy of *something:* the energy of a speeding automobile, the energy of the fuel oil that heats the house. If we are thinking of nothing . . . well then, there is no energy.

But that's not what Richard Feynman learned while a student at college. He was reading up on all the attempts people had made to create a quantum theory of the electromagnetic field— of electricity and magnetism, and of the ripples in the field that are light and X rays and radio waves. What he learned was that nobody could figure out how to make things come out the right way. When even the greatest scientists tried to calculate the energy of *no* electromagnetism, they would never get zero. In fact, they kept getting infinity! Pure nothingness, pure darkness, pure stasis: these did not exist. In their place was a blazing cauldron. The vacuum was an inferno.

Years later, Feynman helped create the theory that was to solve that puzzle. The theory is one of the shining jewels of modern science: he was awarded the Nobel Prize for this achievement. But listen now to what he has to say about his attitude back then toward what he had read—an attitude that I, for one, find astonishing.

I was inspired by the remarks in those books; not by the
parts in which everything was proved and demonstrated
[but by] the remarks about the fact that this doesn't make
any sense. . . . So I had this as a challenge and an inspiration. I also had a personal feeling, that since they didn't
get a satisfactory answer to the problem I wanted to solve,
I didn't have to pay a lot of attention to what they did do.

A college kid had decided that he need not pay attention to
the best efforts of the greatest scientists of the day.

Richard Feynman was the most independent cuss ever to walk
the Earth. He never took anybody's word for anything. He did
everything for himself. While in high school, he decided to rein-
vent trigonometry. Rather than learn the subject from a book, he
set himself the task of deriving all the standard formulas in his
own way. Some of the proofs he discovered turned out to be less
elegant than those in the textbook. But others were better.

Feynman loved nothing so much as a puzzle to be solved,
preferably by some smart and idiosyncratic trick. While a kid
during a summer job as a busboy, he tried to invent a better way
of carrying dishes on trays (everything ended up on the floor).
While working on the atom bomb at Los Alamos during the
Second World War, he had his wife send him letters in a code to
which he did not know the key: he would amuse himself by de-
ciphering them. The whole thing drove the censors crazy, but
he persisted, their discomfort adding zest to the game. Also at
Los Alamos he figured out how to crack safes, and he opened
them at every opportunity—especially those labeled TOP SECRET,
just to terrify the security officers.

Perhaps the ultimate test of Feynman's independence occurred at Alamogordo, New Mexico, at the test firing of the first
atomic bomb. On this occasion, he intentionally exposed himself to the danger of permanent blindness. Everyone had been
issued dark glasses through which to watch the explosion. But
Feynman wanted to see the blast directly. He quickly calculated
that the burst of ultraviolet light could not blind him if he
watched through the windshield of a truck, and he tossed away
the glasses. He later figured he was the only person present to
directly witness the cataclysm.

The general public got to see this gutsy self-reliance in action when he served on the commission investigating the Challenger disaster. On January 28, 1986, the space shuttle Challenger lifted off into a clear blue sky: just over a minute later it exploded in the worst catastrophe of NASA's history. On board were five astronauts, one engineer, and Christa McAuliffe, a schoolteacher chosen in a nationwide competition.

Liftoff occurred after a night so cold that ice had built up on the shuttle. Almost from the start, attention focused on the effects of this cold on the infamous O-rings in the solid rocket boosters. Feynman pursued the issue in his own inimitable way. While everyone else was in Washington conducting press conferences and inconclusive meetings, he was off snooping around on his own: down at the launch site at Cape Kennedy talking to the technicians, or over at the plant that had built the solid rocket boosters, avoiding management like the plague and talking shop with the engineers whose warnings had been ignored. He was often bucking the commission chairman: so contrary did he become that he ended up writing a dissenting appendix to their official report—he called it his "inflamed appendix." It was also characteristic that he found a means of dramatizing the issue in the clearest possible way. Nobody who lived through those times will ever forget his performance one day when, in full view of the television cameras, Feynman dropped a bit of rubber into a glass of ice water and demonstrated that the material quickly lost all resilience. The performance was the high point of the investigation, and it dominated nationwide news coverage for days to come.

What attracts a child to science? It might be a fascination with the wonders of the universe. It might be a respect for the tremendous power of science to effect technological change. My guess is that Richard Feynman was attracted to science because of the latitude it gave him to practice his favorite sport: solving problems in a new way, a smarter way—his way. It mattered not the slightest whether the problem was to develop a quantum theory of electricity and magnetism or to invade the classified safes at Los Alamos. The impulse was the same.

Most scientists spend a lot of time reading up on what other people have done: textbooks to learn a new field, journals to keep up with the latest work. Feynman did neither. He insisted on inventing everything for himself, and so brilliant was he that he usually succeeded. Feynman reinvented the wheel over and over again. Actually, though, that's not quite true. Feynman did not invent wheels. He invented other things, things nobody had ever seen before, but that do what wheels do.

For his Ph.D. thesis at Princeton University, he reinvented quantum mechanics. He had read a few books on the theory and had not understood them. There's nothing unusual about this: nobody truly understands quantum mechanics. On the other hand, most scientists, with the passage of time, learn how to avoid stewing over the mysteries too much. This is what Gamow had done (chapter 3) when he blithely disregarded the theory's enigmatic qualities and went on to use it to account for radioactivity. Feynman's response to the mystery, however, was characteristic of the man: he sat down and entirely recreated quantum mechanics in a new and different form, one that pleased him better than the old.

Such exercises occupy a time-honored position in science. The game is to do the same thing all over again, but differently. The final result you obtain is nothing new, but you got it in a new and interesting way. As a consequence, you have gained an important insight into what the original result meant. Suddenly, it "smells" different.

The best example I know of pertains to Newton's laws of motion. Newton's theory, as presented in his *Principia,* describes the motion of a body as its reaction to the various forces acting upon it. Bodies move because forces push them around. In these terms Newtonian mechanics accounts for the orbits of the planets and the vibrations of violin strings, the swinging of a pendulum and the flow of water. So all-encompassing was this theory, so successful, that it set the tone for all future science.

But about half a century after the publication of the *Principia,* an alternate theory of motion was developed. This new theory made no mention of force. Rather, it dealt with a certain mathematical quantity called "the action." The fundamental law of nature it proposed was that among all possible motions of a body, the one actually followed was that for which the action was the least. It is not hard to show that the consequences of this "principle of least action" are identical to those of Newton's laws. Everything Newton predicts, this new theory predicts as well. So there is no formal reason to choose one theory rather than the other.

But they are not identical in informal terms—in terms of how they make us think about the world. Newton's formulation leads us to think of material objects as passive victims of blind forces; as being helplessly and incessantly shoved this way and that. But by its very nature, the principle of least action irresistibly tugs us in a different direction. It seems to be telling us that a body "tries out" every possible motion, mulling them over in its mind, so to speak, and choosing the easiest. Such a point of view smacks more of free will than helpless passivity, more of a wholistic view of the world as participating in decisions concerning its fate, rather than a victim of blind causation.

Because this alternate theory was not developed until well after the scientific revolution, it never exerted as great a sway over the imagination as did Newton's. But I often catch myself wondering what our view of the universe would have been, had the historical sequence been reversed and the principle of least action been developed first.

For his Ph.D. thesis, Feynman did the same for quantum mechanics. He invented a quantum principle of least action. As before, the formal predictions of this alternate theory were identical to those of orthodox quantum mechanics—but the mathematical details were different, and the informal "smell" was different as well. When Feynman's work was published, few people paid much attention. But by now it has become a staple. Certain problems that were difficult in the original formulation proved surprisingly easy in this new approach. And more than that: Feynman's Ph.D. thesis has altered the very way in which people approach the theory.

While in graduate school at Princeton, Feynman did a few other things as well:

- He stirred heated Jell-O as it cooled, seeking to determine how motion affects coagulation.
- He ferried ants to and fro, seeking to determine to what extent they could communicate with one another.
- He worked on a new (but still nonquantum) theory of electromagnetism, in which the flow of time is scrambled, and cause and effect are partially interchanged.
- He nearly blew himself to kingdom come.

This last episode makes for a good story. Water flowing out of a lawn sprinkler makes it spin. It has to do with Newton's law of equal and opposite reaction: the sprinkler squirts the water out tangentially, and the reaction force from the water sets it spinning in the opposite direction. But around the physics department at Princeton there floated for a time the question—what happens if you reverse the situation? Suppose that instead of squirting the water out, the sprinkler sucks it in. Then which way does it rotate?

It's a nice question, and on the surface it looks simple. But it is not simple. The more you think about it, the more subtle it becomes. The debate at Princeton grew more and more heated. Some people argued for one direction, others for the other. A third group of people changed their minds every few days. Furthermore, as time passed, the argument showed no signs of winding down. Eventually, Feynman decided to settle the question once and for all. He would get some actual data. He would build an "inverse sprinkler" and watch what it did. He snuck into a lab, assembled a crude simulacrum of a sprinkler, immersed it in a big glass bottle filled with water, and used compressed air to power the flow of water backward through the device.

Feynman turned on the air compressor. Nothing much happened. He turned up the power. Still nothing. He gave it one more tweak—and the bottle exploded with a bang. Shards of glass lay everywhere. No one was hurt, but Feynman was banished from the lab.

Sprinklers, ants, Jell-O . . . these were not diversions for Feynman, high jinks to distract his mind before getting down to his true work. They were his true work. They were all science, and science was what Feynman loved and revered. He attributed this attitude to his father. Feynman's father was not a particularly successful man: he was an immigrant who worked for much of his life as sales manager for a uniform company. But it was he who taught Feynman to hate sloppy thinking, and to respect rational inquiry above all else. He would take his son for walks in the woods. They would spot a bird, and the boy would ask its name. His father did not know—and he didn't care. He told his son,

> You can know the name of that bird in all the languages of the world, but when you are finished you'll know absolutely nothing about the world. You'll know about the humans in different places and what they call the bird. So let's look at the bird and see what it's doing—that's what counts.

Feynman relates that as a child he was actually relieved to learn that Santa Claus was a myth. It comforted him to learn that adults could lie, for this was a *simpler hypothesis* than the hypothesis that Santa was able to slide down all those millions upon millions of chimneys at the same time. Conversely, he had a passionate concern for truth, and he wept when he learned that some of the stories in the Bible were made up.

In a revealing episode, Feynman's fiancée once commented that there were two sides to every question: his response was to demonstrate to her the Möbius strip, which has only one side. This simple but striking mathematical construction, made by twisting a strip of paper before gluing the ends together, has no top and no bottom, no inside and no outside. It is a delicious conundrum, and it is what made Feynman love science and mathematics. But at the same time the episode has a darker tinge to it, for his response entirely missed the human truth his fiancée was getting at.

Indeed, Feynman had nothing but contempt for most human endeavor. In college he detested the nonscientific courses they made him take. While his Graduate Record Exam scores in the sciences were off the scale, those in the humanities and social sciences were moronic. "I'm a one-dimensional sort of guy," he was fond of saying, and he was right. My own guess is that his comment on the Möbius strip was no lapse. I think he was intentionally evading the sloppy, nonscientific world of human complexity and ambiguity. In another episode, he confided to a friend his plan for resolving differences with his girlfriend: since he was older and more experienced, the two would agree that his decisions would be accepted.

J. Robert Oppenheimer, director of the scientific effort at Los Alamos that developed the world's first atomic bomb, recounts that a line from the Bhagavad Gita flashed through his mind when the bomb finally went off down at the test site: "I am become death, the destroyer of worlds." It is characteristic of Feynman that at that climactic moment nothing of the sort occurred to him. He was struck by the sudden appearance of clouds above the explosion. Why would that be? Even as the great light of that terrible blast was still dimming, he set about surmising their probable cause.

Feynman's "one-dimensionality" was extreme, and it limited him enormously. But within his chosen dimension, he was a master. With his great and idiosyncratic brilliance, he was perpetually turning established notions on their heads. In a college philosophy class, faced with the necessity of writing a term paper on a subject for which he had nothing but contempt, he hit upon the alternate notion of watching himself fall asleep. For weeks he carefully observed his various states of mind while drifting off, and he wrote them up and turned them in. By these observations, Feynman had characteristically transformed a philosophical question into an empirical one—and it is worth emphasizing that they are both hilarious and fascinating.

Feynman never stopped thinking about the puzzle he had read about while in college: the mystery of the infinite energy of nothingness. It had to do with a quantum theory of electricity and magnetism, technically known as quantum electrodynamics. Throughout his years in graduate school, and afterward during the war at Los Alamos working on the atomic bomb, he kept returning to the problem. He knew that it was no small matter; no minor conundrum, ultimately to be solved by some gimmick. Far from it: everybody knew that it occupied a critical position in the development of physics.

The early decades of twentieth-century physics had witnessed two great revolutions, relativity and quantum mechanics. Once these discoveries were in place, it was natural to extend them in a new direction: to broaden quantum *mechanics* into a quantum *electrodynamics*. But the moment people tried, the vexing infinity arose. Furthermore, the more they thought, the more they realized that the conundrum was fundamental. It was a warning that something was seriously wrong, that the task would be far more difficult than anybody had foreseen. Something new was required, something going far beyond current knowledge. Nobody knew quite what to do.

But in 1947 there occurred a remarkable development. It happened at a historic conference at the Ram's Head Inn on Shelter Island, a small bit of land lying just off the eastern end of Long Island, New York. Gathered were a select group of the country's best physicists: Feynman was one. Three senior scientists gave synoptic lectures—but these were not the high point of the meeting. The high point was a presentation by the young experimentalist Willis Lamb. In an experimental tour de force, Lamb had succeeded in demonstrating unequivocally that the structure of the atom disagreed with the predictions of quantum mechanics.

Nothing is more welcome to a working scientist than the discovery of a gaping hole in established wisdom. Even as they sat there on the overstuffed chairs of the Ram's Head Inn, mulling over Lamb's discovery, everybody felt that it had something to do with the quantum nature of the electrical attraction that binds the atom together. They had returned to a time-worn problem, one beset by seemingly insuperable difficulties—but now Lamb had provided them with something that before they had lacked: a fact. Now they had what Alvarez had loved so much—experimental data, pure, clean, and hard. It was something against which their attempts could now be tested. The time was now ripe for an ambitious effort to push through the difficulties and create a quantum electrodynamics.

Another participant at the meeting was Hans Bethe, a colleague and friend of Feynman. During the train ride home, he did a quick calculation, seeking to understand Lamb's discovery. Bethe's effort could hardly be termed a rigorous theory— but to his surprise he found he was getting pretty much the right answer. So excited was he that he telephoned Feynman during his trip to tell him the news.

When he got back home, Bethe gave a talk on his amazing success, and since Bethe is a scrupulously honest man, he emphasized in the talk not only his theory's virtues but also its problems—the shortcuts he had taken, the unjustified assumptions he had made. When he was finished, Feynman came up to him in the lecture hall. "I know how to do this right," he said.

As indeed he did. Feynman's reformulation of quantum mechanics in graduate school, his work on a time-scrambled electrodynamics there as well, much of what he had done thereafter—everything now came together. Lamb's data gave Feynman his first chance to apply his new methods to a real experimental situation. When he did so he saw with a start that without realizing it at the time, he had created a new way of doing physics. While Bethe had provided many of the essential insights into the correct explanation of Lamb's puzzling discovery, Bethe had relied on earlier ways of doing things, ways poorly fitted to the new physics that was now emerging. But Feynman's strange techniques—idiosyncratic, brilliant— made it all rigorous, and they made it all straightforward. "It was obvious," he later said, "it would work; that was the fun of it. *It would always work.* . . . I began to realize that I already had a powerful instrument; that I was sort of flying over the ground in an airplane."

By the late 1940s the final theory, independently developed by Feynman but simultaneously by others as well, was complete. Quantum electrodynamics is one of the shining jewels of physical science. It has proved enormously successful: indeed, at present it is the most accurate scientific theory that has ever been created. As an example, the experimentally observed value of a measure of the magnetic field of the electron is

$$1.001\ 159\ 652\ 5$$

while the theory's prediction for this quantity is

$$1.001\ 159\ 652\ 2$$

The accuracy of the theory is equivalent to measuring the distance from New York to Seattle to a hundredth of an inch.

Quantum electrodynamics—QED for short—has spilled over into other areas as well. The theory is more than accurate: it is fecund. It has turned out to be pregnant with many children. Its techniques have proved enormously fruitful in unrelated areas as well: in the physics of the solid state, for instance, and in that of elementary particles.

Perhaps most remarkable of all is that this great triumph was achieved in the face of a deeply disturbing—indeed an ominous—difficulty. Quantum electrodynamics, when applied to a large number of actual problems, turns out to make predictions that on the surface appear to be quite nonsensical. Its predictions are usually infinite.

The situation is reminiscent of the infinity that had worried Feynman as an undergraduate. His theory's prediction for the Lamb shift, for instance, was infinite. How could such a bizarre result be understood? The method people developed to deal with this conundrum is brilliant, and it is worth describing.

It can best be understood with the help of an analogy. Leaves tumbling from a tree in autumn drop at different speeds. They do so because of air resistance. A flat leaf scoops up a lot of air, and so falls slowly; the same leaf crumpled into a ball scoops up less, and so falls faster. Suppose, however, that we wish to know how fast the leaf "really" falls—falls in the absence of air.

In this analogy, the rate of fall of the leaf corresponds to various properties of the electron we might be interested in—Lamb's result is one of them—and the resistance of the air corresponds to the incessant action of the electromagnetic field, which messes things up. The analogy, though, is incomplete. It fails in two ways. The first is that in the case of the leaf we can always get rid of the disturbing element, by dropping the leaf in

a vacuum. But there is never any way to get rid of the electromagnetic field. And the second failure is that the effect of this
field is infinitely great.

QED deals with this as follows. People realized that in calculating Lamb's phenomenon, they were not merely calculating
the theory's predictions for this one effect alone. They were
calculating *all* the things the electromagnetic field did to the
electron. Only one of these was what Lamb had measured.
But there were others. One turned out to pertain to a shift in
the electron's mass. Another pertained to its charge. So they
could subtract these extraneous effects from the total. The
remainder pertained solely to what Lamb had studied—and
this turned out to be not only finite, but in perfect accord with
measurement.

It is like discovering that leaves "really" would drop infinitely
rapidly, but that the air exerts upon them a retarding force that
is also infinite, and that slows them down to a gentle fall. To a
degree this subtraction procedure is a wonderful triumph. But
there is an ominous tinge about the whole situation. It is, indeed, disquieting to think that the electron's "true" mass and
charge are infinitely great. Nobody would be disturbed if the
corrections involved in this subtraction procedure were finite.
But infinity is a dangerous thing to play with. Indeed, although
infinity may be a magnitude in some sense, it is not a number
in any ordinary meaning of the term. Deadly paradoxes arise if
we try to treat it as one.

The most unsettling such paradox I know of pertains to an
imaginary hotel with an infinite number of rooms. Suppose
they are all taken. The infinite hotel is fully occupied—and now
I show up at the front desk, asking to be admitted. A normal hotel would turn me away, but the manager of the infinite hotel
has a trick up his sleeve. He knows a way to get me in.

The manager simply puts me into room number one. There's
a doubling up, of course. To solve the problem, to clear out
space, the manager then moves the original occupant of room
number one into room number two. To prevent a doubling up
there, its original occupant is then moved into room number
three. And to prevent a doubling up in the third room, its original occupant is moved into. . . .

There is, of course, no end to this process. A finite hotel eventually runs out of rooms to accommodate the displaced guests. But the infinite hotel does not. It always has more space. And so, even though it was fully occupied when I walked in the door, it has room for me. Furthermore, the same argument applies to anyone else—to you, say, and to all your friends. There's room for everyone. So, was the infinite hotel full after all?

The difficulties entailed by QED's subtraction procedure illustrate a profound advantage that science enjoys over disciplines such as mathematics and philosophy, which do not have the benefit of experimental input. Hard data can sometimes help us deal with otherwise intractable paradoxes. Feynman had Lamb's results against which to compare his calculations. No one can be comfortable with infinite subtractions, but the beautiful agreement of the final result with experiment is persuasive evidence that for all its flaws, something is deeply right about the theory. Nobody quite knows where the solution to this conundrum will lie. But most people are convinced that when it is ultimately found, the solution will preserve QED's wonderful successes.

The quantum theory of electromagnetism is a hideously complex maze of mathematics. The student wishing to learn the subject must be prepared to spend many a month wading through some of the most daunting formalism in all of science. The book from which I myself studied QED ran to 905 pages, each densely packed with unfamiliar symbols, formulae, equations, graphs, and footnotes; all in mighty small print. The thing weighed a ton.

Much of that formalism had been developed by Feynman. He was a master of the abstruse mathematics required by the theory. Many of the special tricks, the short cuts, the detailed techniques needed to actually perform its calculations were invented by him. But he had also invented something else—Feynman diagrams.

These wonderfully simple, vivid sketches cut through all the maze of mathematics, and they give a simple, intuitive feeling for what is going on. Each diagram is a trivial affair, reminiscent of the stick figures children draw—a mere few lines, some straight, some wiggly, connected into Vs and loops. Accompanying each is a description of what it represents: such and such a diagram illustrates an electron that emits a photon, which itself travels along, temporarily splitting into a subsidiary electron and a positron before recombining and then being absorbed by a second electron.

The fascinating thing about Feynman diagrams is that this is not what they really mean at all. Actually, each is an instruction to carry out a certain mathematical operation. Every problem in QED requires one to do a long list of calculations, each quite difficult. Even to *find out* which calculations need to be done is no mean feat—and it was for this purpose that Feynman invented his diagrams. Every element in them has a well-defined, purely calculational meaning. A straight line is an instruction to write down a certain mathematical quantity: the junction of a straight and wiggly line another instruction to multiply this by a second quantity. Feynman had invented these diagrams as a sort of mnemonic aid, a bookkeeping device that summarizes for us these complex operations. They are a means of ordering in our minds the work before us.

He had taken the complex mathematics of the quantum theory of electromagnetism and done an astonishing thing: he had created a second level of analysis; a new kind of intuition, contained in the verbal description accompanying each diagram, which enables us to think about QED processes in a simple, intuitive way. But nothing about the diagrams was informal or sloppy. They were bound by rigorous, well-defined rules of construction. And while on the one hand, these rules were essentially visual in nature, on the other hand they corresponded exactly to the mathematical operations they were designed to illuminate. I know of nothing else quite like the Feynman diagrams in all of science: they form a separate intuition, over and above the normal understanding provided by QED; one that simplifies and illuminates the fundamental theory that he himself had created.

Feynman was not only one of the great mathematicians of his day. He was also one of the great intuitionists. He constantly strove for an understanding that could be expressed in simple, down-to-earth terms. Nobody else could have invented these diagrams.

One day a colleague of Feynman's asked him a technical question. "I'll prepare a freshman lecture on it," Feynman replied. Several days later, however, he was forced to admit defeat. "I couldn't do it," he said. "I couldn't reduce it to the freshman level. That means we don't really understand it."

It is characteristic of Feynman that he said that *we* did not understand the phenomenon, rather than that *he* did not. We are, after all, speaking here of one of the great egos of our age. But it is also characteristic that he sought to reduce the question to an elementary level. Only when he could express something in such terms, Feynman felt, did he truly understand it. As a consequence, he was one of the great educators of our time.

By the time he reached middle age, Feynman had become one of the most famous scientists of his generation. He used a secretary to ward off unwelcome visitors. But this policy was never extended to college students. His door was always open to them. For many years he gave a course entitled Physics X, from which all but undergraduates were rigorously excluded. Physics X had no homework problems and no exams; it gave no grades and offered no formal credit. It did not even have a well-defined topic. The subject matter was whatever the students felt like discussing. Many alumni now recall it as a high point of their undergraduate education.

Feynman was in the habit of regularly driving over to a local high school to meet with physics classes. After his Nobel Prize, the sessions became clogged with reporters and the like, with whom he had no desire to consort. Rather than cancel the visits, he asked the school administration to keep them secret. Feynman wanted nothing so much as to talk physics, and he wanted to talk it only with the kids.

By the early 1960s the introductory physics course at Cal Tech, where Feynman taught, had grown stale. It was being primarily taught by teaching assistants who were themselves only a few years older than the freshmen. Nobody ever should have expected them to do a good job of it, and they did not. In the middle of the year, a delegation of freshmen took their complaints to the department head, who responded by asking Feynman to lend a hand by conducting a weekly question-and-answer session.

One of those teaching assistants, now a mature scientist, vividly recalls the difficulties he experienced dealing with some of the brightest students in the country. Time and again they would stump him with smart questions. At one point a student proposed what appeared to be an insoluble paradox that demolished in an instant the theory of relativity. The teaching assistant could not figure out what to make of it. Was one of the great revolutions in thought of our century in danger of collapse? He temporized, and recommended the freshman take it to Feynman's session.

"Ah, yes," Feynman responded, when the freshman challenged him with the paradox. "There are lots of paradoxes like that. They're all trivial." He then stood silently in the lecture room, gazed at the ceiling, and thought.

Time passed. Several hundred students, forgotten, sat silently. Feynman continued to think. More time passed, and the students began to shift uneasily in their seats. After what seemed like an eternity, Feynman suddenly came to life, and he sketched on the blackboard the resolution of the paradox.

"So there you see," he told the assembled multitude, "it *is* trivial."

Feynman's experience that year persuaded him that the education Cal Tech was giving its freshmen was all wrong. He resolved to take over the introductory course himself. The result may not have been quite what the students wanted, but it led to one of the most remarkable educational ventures of all time. For two years, he all but abandoned research, and he threw himself into the task of reducing all of physics to its most essential terms. In this ambitious project he worked eight to sixteen hours a day, five days a week. The upshot for each class meeting was summarized on a brief scrap of paper, upon which he had jotted a few laconic notes—but the lectures he gave to the astonished freshmen were like nothing they had ever experienced before.

A further product of the course was a monumental three-volume textbook, *The Feynman Lectures on Physics*. It has by now attained the status of a classic. I know of no other book like it, either in physics or in any other field. In this extraordinary text, Feynman entirely recreates all of his discipline from the ground up. In the opening chapter, he asks perhaps the most ambitious question a scientist might put: what is the single most important thing that science has taught us? His answer is the proposition that Ludwig Boltzmann spent his life fighting for: that everything is made of atoms. He explores this in a discussion of remarkable scope. In *a single printed page* he shows how the atomic hypothesis explains (1) why liquids change their shape but not their volume when flowing, (2) why the pressure of a gas is proportional to its density, (3) why compressing a gas heats it, and (4) why ice expands upon cooling. His discussion throughout this section is entirely qualitative, and it employs not a single equation or mathematical symbol.

But the mathematics soon begins. Within a matter of weeks the students were making their way through Einstein's theory of relativity. I first encountered this subject while a sophomore: these students were freshmen. But the discussion Feynman gave was at a depth and level of sophistication more appropriate to a senior-level course. Within a few more weeks, the level of sophistication escalated yet again, and in discussing "the Origin of the Refractive Index," the hapless students were dealing with material at the graduate level.

It cannot be said that for its intended audience the course was a success. It demanded more intellectual maturity, a deeper understanding of physics, than most freshmen could be expected to possess. Bit by bit, as the weeks rolled by and the lectures soared to ever higher reaches, more and more of the students stopped coming to class. But the classroom remained full. Greyer hairs were cropping up here and there among the "real" students. The faculty at Cal Tech—physicists and electrical engineers, postdoctoral fellows and full professors—were sitting in on the lectures. They had gathered to relearn their trade.

It is not clear whether Feynman ever paid much attention to the transformation.

Indeed, there is no use thinking about *The Feynman Lectures on Physics* in terms of an audience of beginning college students. With the passage of time, the real audience that has developed for this work has proved to be—other physicists. In chapter 2 of this book, I described Boltzmann's work on entropy and its connection with the one-way flow of time. In preparation for that chapter, I had spent much effort going over Boltzmann's work. But I also did something else. I looked up what Feynman had to say about the subject to his freshmen. For I knew I would find there the essential intuition, the underlying insight that reduced this difficult subject to its most comprehensible form. What I found surprised me at first. It was a treatment of a subject that at first sight seemed unrelated: the ratchet and pawl, the commonplace mechanical gadget that rotates one way but not the other. Feynman had hit upon the notion of embedding one in a gas, and he used it to illustrate the subject of entropy in the simplest possible terms. In contrast to Boltzmann's approach, which was deeply mathematical and abstract, I found Feynman's breathtaking in its simplicity and clarity: in all the literature of entropy, I know of nothing quite like it. Such high points are common throughout the book.

One day, to their immeasurable horror, a group of physicists found themselves attending a conference at a country club of truly grotesque opulence. That evening everybody stood around complaining about the place—everyone but Feynman. He simply grabbed his suitcase, headed for the door, and marched off into the woods. The next morning he reappeared, somewhat disheveled but none the worse for wear. He was sixty-two years old at the time.

It's a great story. Here's another. One day Feynman was having lunch with a graduate student who told him he was planning to found a company that would build the world's first massively parallel computer. "That is positively the dopiest idea I ever heard," Feynman responded, and by the end of the lunch he had signed on with the fledgling firm as a summer employee. When Feynman reported for work and asked his new boss what to do, he was told to go off and think about the new machine's possible applications. But Feynman would have none of it. "That sounds like a bunch of baloney," he responded. So they sent the Nobel Prize–winning full professor off to buy office furniture.

Stories about Richard Feynman are so much fun that by now they have attained the status of a minor art form. In yet a third, Feynman is sitting at his kitchen table working, and he has just made an important discovery. As he relates it, here is what happens next:

> After working some more, it got to be very late at night, and I was hungry. I walked up the main street to a little restaurant five or ten blocks away, as I had often done before, late at night.
>
> On earlier occasions I was often stopped by the police, because I would be walking along, thinking, and then I'd stop—sometimes an idea comes that's difficult enough that you can't keep walking; you have to make sure of something. So I'd stop, and sometimes I'd hold my hands out in the air, saying to myself "The distance between these is that way, and then this would turn over *this* way. . . ."
>
> I'd be moving my hands, standing in the street, when the police would come: "What is your name? Where do you live? What are you doing?"
>
> "Oh! I was thinking. I'm sorry; I live here, and go often to the restaurant." After a bit they knew who it was, and they didn't stop me any more.
>
> So I went to the restaurant, and while I'm eating I'm so excited that I tell a lady that I just made a discovery. She starts in: she's the wife of a fireman, or forester, or something. She's very lonely—all this stuff that I'm not interested in. So *that* happens.

Feynman spoke loudly, and he had a New York accent so thick you could cut it with a knife (he had been born and raised in Queens). When he talked about physics, he would wave his hands, stride back and forth, burst out with jokes. Gales of laughter would erupt from the rooms in which he taught his abstract and difficult courses. Friends whom he regaled with his new ideas often found the experience not just interesting, but sidesplitting as well. There was a time when people worried that he might be depressed. But it was hard to be sure: how could you tell the difference between Feynman depressed and anybody else at his most exuberant?

In a photo taken at a university talent show, Feynman plays the drums. Surrounded by students, all properly deferential and carefully dressed, he is the only man present not wearing a tie. The students lean forward in respectful, attentive poses: Feynman, in contrast, is exhilarated, one arm thrown upward and face turned ecstatically toward the ceiling. Another photo (page 116) shows him in conversation with P. A. M. Dirac, Homi Bhabha's old mentor (chapter 4). Dirac is tall, gaunt, willowy: leaning back languidly against an urn, he looks like a romantic poet slowly wasting away of consumption. Feynman on the other hand is leaning forward, and he is gesturing excitedly. He looks like a used car salesman just about to nail down a deal. In a third photograph, taken at the Nobel Prize awards ceremony in Stockholm, he is shown at a formal dinner seated next to a princess. Uncharacteristically, Feynman is done up in a tux; not so uncharacteristically, a cigarette dangles loutishly from his lips. He is squinting nastily at the camera.

Feynman kept a coat and tie in his office, but he made a point of "forgetting" to wear it to the elegant lunchroom at Cal Tech. In any event, he often preferred eating in the student cafeteria, just to keep an eye on the young women. Well into his thirties, he was in the habit of attending student dances and trying to pass himself off as one of them. He played bongo drums at parties, whether anyone asked him to or not. He boasted that he could move one eye and not the other. He would challenge people at lunch to set him a problem that could be stated in ten seconds that he could not solve to 10 percent accuracy within a minute. He showed up at parties and pretended to get drunk (in fact he avoided alcohol). He showed up at one party dressed up as God.

Richard Feynman was incessantly clowning and strutting and boasting. If you liked this sort of thing, he was a delight: if you did not, he could be intolerable. Undergraduates loved him—one of the world's great scientists, goofing off like a kid. As for his colleagues, sometimes they were not so sure. A friend once described him as half genius and half buffoon. But after getting to know him better, he revised his judgment—to all genius and all buffoon.

The clowning was an act, of course. Far from being a loutish showoff, Feynman was in truth a deeply private person. Indeed, he reacted angrily when anybody tried to pierce his facade.

Nowhere is this reticence more evident than in his famous book of recollections, *Surely You're Joking, Mr. Feynman!* The *first mention* of his wife appears on page 104 of this book, and it consists merely in the statement that she was in a hospital. The reader might be interested to know what was going on, but that is not what Feynman is about in this chapter. Rather, he is describing how he was able to tell by smelling it if anybody had recently handled an object. Three pages are devoted to this nonsense. Nothing about this passage—or any other passage in the book—gives the slightest impression of the truth about his marriage, of its tragic conclusion, and the devastating effect it had on his life.

Feynman had met Arline Greenbaum as a teenager. They quickly became sweethearts, and in college engaged to be married. But shortly thereafter she developed a lump on her neck. A severe fever followed: she entered a hospital. With the passage of time, Arline was in and out of hospitals, and the symptoms proved difficult to diagnose. Eventually, her doctors realized that she was afflicted with tuberculosis of the lymphatic system. The disease was incurable, and it was fatal.

Feynman, now a graduate student, was told. He and Arline had promised to be completely truthful with one another, and he felt bound to tell her in turn. But everyone else felt that such a death sentence would be too devastating for her to bear. Her doctors, her parents, his parents—even his kid sister—everyone prevailed upon him to lie to her. Miserable, standing by her bedside at the hospital with her parents beside him, he told her she had "glandular fever." But not long thereafter, home from the hospital, Arline heard her mother weeping. She confronted Feynman. He told her the truth and proposed that they marry immediately. She accepted.

But it could not be. Their decision unleashed a storm of protest from his parents. They were terrified that he would inevitably contract tuberculosis as well. His mother argued vehemently with him, accusing him of throwing away his life. At the same time, Feynman learned that the graduate fellowship he held at Princeton was exclusively reserved for unmarried students. The department refused to consider bending the rules on his behalf. His parents were not wealthy and could not afford to support him in school: to marry would be to give up his career. So they did not marry.

Arline grew steadily worse. Most of her life was spent in hospitals. He would visit her whenever he could. Eventually, he obtained his graduate degree and was offered a job. Able at long last to support a wife, he borrowed a car immediately after graduating and outfitted it with a mattress. He picked her up from her parents' house, and they took the ferry over to Staten Island. They were married by a justice of the peace. Their honeymoon was the ferry ride back. She went right back into the hospital again.

Within months, Feynman was out at Los Alamos working on the atomic bomb, and Arline had been transferred to a hospital in Albuquerque. He did not have a car, and would borrow one to visit her. Sometimes he was forced to hitchhike all the way to see her on weekends. Outside the hospital, he would barbecue steaks on a grill while the traffic roared by on Route 66. Weekdays, when he was back in Los Alamos, the couple would write one another, sometimes as often as once a day: letters suffused with love and mutual solicitation, simultaneously comforting and kidding one another, and passing along the news. They had not yet lived together. They had not yet made love. Ultimately, with her weight down to eighty-four pounds, they decided to consummate their marriage (Arline made sure no one would invade her hospital room). Shortly thereafter, in celebration of her husband's birthday, Arline from her deathbed arranged for the entire Los Alamos laboratory to be flooded with fake newspapers bearing the headline "Entire Nation Celebrates Birth of R. P. Feynman."

One day at work Feynman was phoned with the news that his wife's death was imminent. He borrowed a car for his last drive to see her. It developed a flat tire, then another and finally a third. Ultimately he abandoned it in desperation, and set off to hitchhike the remaining distance to the hospital. A few hours after he arrived, his wife lay dead.

In *Surely You're Joking, Mr. Feynman!* Feynman describes briefly his wife's death. I find the account he gives there unnerving— not for what he says, but for what he does not say. After baldly stating that she had died, he immediately changes the subject to the curious behavior of a mechanical clock by her bedside. After a good deal of this, he returns to Arline, but merely to comment that her death had left him curiously unmoved. And aside from a brief return to the subject far later in the book, this is all he has to say about her.

Driving back to Los Alamos, he refused to allow anyone even to speak of her, let alone comfort him. Soon after, back in Far Rockaway, he telephoned an old friend with the proposal that they head down to the boardwalk and try to pick up some girls. So far as the friend could tell, Feynman was utterly serious. At the time, his wife had been dead for precisely three days.

And yet, to the end of his life Feynman never ceased grieving for Arline. Two years after her death, he wrote to her, in a letter so filled with heartbreak and loneliness and grief that it brought tears to my eyes when I read it. He folded this letter in an envelope and preserved it carefully for the rest of his life. In later years he would on rare occasions bring out an ancient suitcase, filled with a few of her belongings, and gaze at them with deep emotion. More than forty years after her death, as he himself lay dying, he wept uncontrollably for his greatest love.

But for all their pathos, there is an element of unreality to Feynman's feelings about Arline in his later life. He clearly never resolved his grief over her. The story of their marriage and her death is tragic and deeply moving—but many people die young, and after a period of mourning, their spouses usually manage somehow to get on with their lives. In some essential way, however, Feynman did not. Perhaps he never stopped mourning her because he never truly mourned her.

Feynman married twice more: the last seems to have been a happy relationship. But for decades after Arline's death, he was unable to achieve a balanced relationship with women. Rather, he oscillated wildly between two radically opposed extremes. On the one extreme was Arline: on the other, everyone else. In these years Richard Feynman slept with every woman he could get his hands on. He studied the ways of predatory women at bars and sought to beat them at their own game. He slept with whores. He slept with the wives of colleagues and friends. He slept with the wives of his graduate students and postdoctoral fellows, men dependent on Feynman for their livelihoods and professional advancement. When he traveled to meetings, his hosts felt obligated to provide him with introductions to suitable, good-looking candidates for his adventures. There is not the slightest reason to believe that he entertained any love for these women. He would save their love letters to him: in one, a woman pregnant by him corrects his spelling of her name.

At the other extreme from these cold, loveless couplings was his tragic love for Arline, and the grief for her that never ceased. The story of this love is one that brings tears to one's eyes—but it is not the story of a normal relationship. Entirely missing from their love were the sort of prosaic, day-to-day activities that occupy most marriages: diapering the baby, balancing the checkbook, making sure the car gets to the mechanic on time. But Richard and Arline Feynman never lived together as man and wife. That is not a very good introduction to the pleasures and responsibilities of marriage.

In thinking about the enormous contrast between Feynman's brutal and promiscuous relationships with most women, and the storybook quality of his first love, I often catch myself wondering what would have happened had Arline lived. Would their marriage have survived? Sometimes I think not.

Feynman could be cruel. When visiting scientists came to deliver lectures on their work, he would invariably occupy center seat in the front row, and he would taunt and hector anyone he felt did not measure up. Once he so humiliated a world-famous but aging visitor that a younger colleague left the room in disgust. He refused to accept the multitudinous and taxing responsibilities that others shoulder as a matter of course: he would leave for the summer having failed to turn in grades for his courses, he would promise to write letters of recommendation for younger people seeking jobs but fail to do so, he never served on university committees or helped write grant proposals. He was often harsh and cutting in his relations with his fellows. A colleague of Feynman—a colleague, by the way, who himself has been awarded a Nobel Prize—recalls that most of his fellow professors were afraid of him. He added that so far as he knew, no one on the faculty ever got close to Feynman.

His brusqueness and lack of concern for the feelings of others extended to his research students as well. One Monday morning he told a student that over the weekend he had solved the problem that he had suggested as the student's Ph.D. thesis research. The student had spent half a year struggling with the subject: he promptly left Feynman's tutelage and transferred to another area of physics. A second former student describes Feynman's attitude toward his research students as "competitive." He felt sure that Feynman had not the slightest interest in their welfare. Any personal problems they might have, their morale while wrestling with their first taste of research, how they were faring in their searches for employment after graduating—these interested him not in the slightest, and he would brush away their efforts to discuss them. It was the physics that interested him.

It was the physics that interested him—and where the physics is concerned, there are other stories to tell, stories of scrupulous honesty and a fine generosity of spirit. Yet a third former student has described how he had worked with Feynman for years to reach an understanding of a certain phenomenon, and how they ending up creating a theory that agreed beautifully with experiment. When the student was finished he wrote up the results for his Ph.D. thesis. Feynman read the thesis with the utmost care—and he found a subtle mistake that the two of them had made. Feynman's correction turned out to bring the theory they had worked so hard to develop *out* of agreement with experiment.

If a piece of his work failed in such a way it might be good enough for a student's degree, good enough for publication, but in Feynman's eyes it had failed in an essential respect. He had not an ounce of possessiveness toward the fruits of his labors. After he had developed QED, he learned that a friend had published a paper interpreting his work in a new light. Feynman himself had not yet written up this work, so that it was his friend rather than he himself who was publishing his creation. But Feynman showed not the slightest irritation, commenting merely "Great! Finally I am respectable." On another occasion he learned that a young colleague who had not yet established his reputation had done a piece of work that duplicated something he, too, had done: Feynman tucked away his research, never publishing it, so allowing the younger physicist to claim the credit.

When Feynman died, a book of reminiscences of him by other physicists was brought out. It is studded with terms such as "brilliant," "creative," and "the most inventive of his generation." That is a signal honor, and it is testimony to the immense regard in which he was held by members of his own profession. Indeed, he was one of the two highest-paid members of the Cal Tech faculty. It is hard to name another scientist who so commanded the admiration and respect of his colleagues.

I am not surprised. Feynman's work was stunning in its brilliance, and it was utterly unique. Everything he did bore his own special stamp: like a Picasso, there is no mistaking a Feynman.

Perhaps the best example I know of pertains to his work on superfluid helium. If the helium in a balloon is cooled down to close to absolute zero, it first liquifies and then, when exceedingly cold, makes a transition to a mysterious state known as superfluidity. Superfluid helium is astonishing stuff. A closed container of it, if given a stir, will continue circulating for months. Alternatively, if some is poured into an open glass its level will smoothly drop until, within a matter of minutes, the glass has emptied. The stuff has invisibly crept up the walls, over the lip, and so out onto the table. Honey has a large viscosity, water a lower viscosity: superfluid helium has no viscosity at all.

Some years after his work on QED, Feynman asked whether these strange properties might be caused by quantum mechanics. Quantum theory was developed to account for the behavior of submicroscopic objects such as electrons, atoms and the like, and in normal circumstances it has little to say about the workings of the large-scale world. Nevertheless, since the early days, people had suspected that superfluidity was a quantum manifestation. The question was how. With the passage of time several theories were proposed: all of them were partial and incomplete.

They were partial and incomplete because everybody knew that the full quantum-mechanical theory of superfluidity would be impossibly difficult to work out. The equations of quantum mechanics grow harder and harder to solve as they are applied to greater and greater numbers of particles. A single particle turns out to be easy; two are hard but the job can be done. But so difficult is the task, that no exact results have ever been obtained for a mere three particles. And to work out the exact theory of superfluid helium would require solving some 1,000,000,000,000,000,000,000,000 equations, each dauntingly complex.

No one could possibly hope to do such a thing—not even Richard Feynman. But what he *did* do was unique, and it was spectacular in its audacity and brilliance. Abandoning any attempt to work out the mathematics exactly, Feynman asked an astonishing question: what general properties must the solution possess, if it is to be in accord with the principles of quantum theory? And then he went on to answer this question in general, intuitive terms; and to show how his answers explained superfluid helium's strange properties. Even today, rereading these papers decades after I first encountered them is an extraordinary experience. Feynman's work in these pages is breathtaking.

This is what Mark Kac, a mathematician of no mean ability himself, has to say about Richard Feynman:

> There are two kinds of geniuses, the "ordinary" and the "magicians." An ordinary genius is a fellow that you and I would be just as good as, if we were only many times better. There is no mystery as to how his mind works. Once we understand what they have done, we feel certain that we, too, could have done it. It is different with the magicians. . . . The working of their minds is for all intents and purposes incomprehensible. Even after we understand what they have done, the process by which they have done it is completely dark. They seldom, if ever, have students because they cannot be emulated and it must be terribly frustrating for a brilliant young mind to cope with the mysterious ways in which the magician's mind works. Richard Feynman is a magician of the highest caliber.

Richard Feynman is probably the most famous physicist of our time. His popular books, his service on the Challenger commission, innumerable TV specials—these have carried his image to a wide audience. But this image, the image the general public has of him, is not the real Feynman. His two books of personal reminiscences are wonderful works—the only books by a scientist I have ever encountered that regularly set me into peals of laughter as I read them. But they, too, are not the real Feynman.

Feynman was a compulsive showman, and he worked hard to embellish his reputation. That reputation was of the wise-guy kid, the perpetual adolescent, the buffoon. But all this was an act, a mask he carefully wore before the outside world. When he was doing physics, on the other hand, all this act dropped away, and an utterly different man emerged. This was a man of the utmost seriousness, passionately devoted to his craft; a man who kept detailed logbooks, carefully cross-referenced, of each day's work; a man who spoke like a gangster, but whose words were of a burnished perfection; a man whose scientific work was Mozartean in its purity and grace.

Indeed, the comparison between Feynman and Mozart is apt. Feynman's scientific work has the same brilliance, the same polished elegance and exquisite perfection as Mozart's music. And as depicted in the film *Amadeus,* Mozart, like Feynman, was perpetually clowning and strutting when not working at his craft. The major difference between the two men lies in the public perception of them. For Feynman's work was technical and specialized, and most people have no access to it: only physicists know the real Richard Feynman. But anybody can appreciate Mozart's music.

It is an unfortunate situation, and one I would like to reverse. I want to show Feynman doing what he honored more than all other things, and what he did with such electrifying skill. I want to close this profile by giving the reader a glimpse of his true self, by reprinting a selection from one of his scientific papers. The passage I have chosen is the closing of perhaps his most important work on superfluid helium. In looking over this excerpt, I am sure the reader will wish to understand it. Unfortunately, lacking years of study, this will be impossible. But I would counsel that it is not necessary to understand the passage in order to sense the magic and the beauty of it. After all, we don't need to know Latin to listen to Bach's B Minor Mass. Indeed, I would recommend that the reader seek not so much to *understand* the passage as to *feel* it, to feel the music of it.

In this passage, Feynman is seeking to understand the means whereby helium loses its superfluidity when heated. He writes,

Of the following I am not sure, but it does seem to be an interesting possibility: For a long line there are an enormous number of shapes and orientations available. Such a line is not infinitely flexible, of course, for the curvature cannot well exceed 1/a. It may be likened to a chain of a finite number of links. Adding one link requires an energy E, say, of order D, but increases the number of orientations by some factor, asymptotically say s. In equilibrium, then, the number of chains of n+1 links is a factor s exp (−E/kT) times the number with n links. For low temperatures this is less than unity. No long chains are important. The excitations consist of rotons and a few other rings of slightly larger size. As the temperature rises, however, there comes a time when the factor s exp (−E/kT) exceeds unity. Then suddenly the rings of very largest length are of importance. . . . At first the curve doesn't make full use of all of its orientations and higher entropy. But as the temperature rises a little more it squeezes into the last corners and pockets of superfluid until it has no more degrees of flexibility available. The specific heat curve drops off from the transition to a smooth curve and the memory of the possibility the Helium can exhibit quantum properties in a unique way is lost.

These are the words of a master.

Richard Feynman died in 1988, after a decade-long battle with cancer. He kept on doing science right up to the end. At one point someone had asked him how much he worked each week. He responded that he really couldn't tell, because he never knew when he was working and when he was playing.

7

Big Science

Martin Perl and the World
of High Energy Physics

When Martin Perl was seventeen, he wanted to fight in the Second World War. Being underage, they wouldn't let him in. So he joined the merchant marine instead. About a year later, an officer found him reading Proust one day while on duty. The man told him to put the book away or suffer the consequences. Perl kept on reading and was kicked out of the merchant marine—promptly to be drafted by the army, so getting his wish.

He got into physics in more or less the same roundabout way. Initially, Perl was a chemical engineer, but his wife felt physics would make for a more interesting career. Partly due to her urging, partly due to an inspiring teacher, Perl enrolled in graduate school in physics after several years working in industry. By now he has become one of the field's most eminent practitioners—he won the Nobel Prize for physics in 1995.

Right now, Martin Perl is showing me around SLAC, the Stanford Linear Accelerator Center. Like the geologist's hammer or the biologist's microscope, SLAC is a scientific instrument. But unlike a hammer or a microscope, this particular instrument is 2 miles long. It has a staff of fourteen hundred people, and it runs through a budget of $140 million each year. When it is on, the beast consumes more than 20 million watts of power, which are brought to it through its own set of power lines and cost several million dollars each year. High-energy physics is but a single branch of physics among many, but SLAC utterly dwarfs any university physics department. Indeed, administratively it is not a department at all, but a separate school at Stanford, like the law school or the medical school. Down there on the main campus, they do small science. SLAC does big science.

Had the Superconducting Super Collider been built, the behemoth would have represented yet another step upward: not big science but megascience. The emergence of such gigantic enterprises is one of the most striking transformations of science in our time. Martin Perl's career spans this transformation in its entirety. He finds the field simultaneously exhilarating and frustrating. On the one hand, he is one of its most vocal critics. At the same time, he is one of its masters.

SLAC accelerates particles to enormous energies, slams them into one another, and studies the outcome. Perl had arrived there in the early 1960s. Almost immediately upon arriving, something about the accelerator caught his attention. The particles that SLAC accelerated were electrons. But these electrons could be made to strike a target, and when they did so, Bhabha's muons were produced. SLAC made muons, lots of them—and muons were interesting.

They were interesting in ways that Bhabha had not realized, that had become apparent only relatively recently. On the one hand, muons were remarkably similar to electrons. But on the other hand, they differed radically from them. As for the differences, one was that the muon was a good deal more massive than the electron, 208 times more massive in fact: that was the comparison between an automobile and a baby. But this was not the only difference. Were the muon nothing more than a heavy electron, it could decay into an electron with an accompanying burst of light. But this decay was never observed. Some further attribute apparently separated the two particles, preventing their direct interconversion. But of what did this attribute consist? People had not the slightest idea.

The truly surprising thing was that, in every other regard, electrons and muons were identical. For more than a decade, a major thrust of Perl's work at SLAC was the search for further differences between these two particles. In experiment after experiment, he probed to find them. He never found any.

Furthermore these two particles, alike in so many ways, were alike in ways that differentiated them from all the rest of the subatomic world. On the one hand, electrons and muons were elementary: they were fundamental units of which matter was composed. Most other particles, in contrast, were not elementary, but were composed of yet smaller particles, the quarks. And on the other hand, so far as either he or anybody else could tell, electrons and muons had no size. They had no extent in space at all. Other particles, however, were extended, somewhat fluffy structures, occupying definite volumes of space. In all the subatomic roster, only the quarks resembled electrons and muons in being truly elementary, pointlike objects. But quarks differed from electrons and muons in a number of other ways— in their electric charges, for instance, and in their avoidance of the strong force through which the quarks interacted.

It was altogether a curious situation: two particles, clearly set apart from all the others, almost but not quite identical to one another. The place of the muon in the roster of subatomic particles was utterly unclear. Was it not extravagant of nature, unnecessarily luxurious, to make room for the muon? What was the nature of the invisible attribute that distinguished it from the electron? What was the significance of its greater mass? And why was there only one such "heavy electron"? Why were there not yet others? Perl found himself continually turning these questions over in his mind—turning them over, and finding no answers.

Finally, early in 1973, there occurred a major creative leap in his thinking. Frustrated by his continuing failures to find differences between the electron and the muon, he decided to ask a different question. He decided to ask if the muon was alone. Was there really only one such "heavy electron"? Or could there be yet another: not a heavy electron, but a heavier electron? Such a particle, if it could be found, might shed light on the electron/muon mystery. Perl decided to search for it.

But how could he search for an unknown particle? How was he to find an object of whose very nature he knew nothing? In the world of high-energy physics, you find things by creating them—creating them in the debris of a collision between two normal particles that have been whipped up to enormous energies in an accelerator. Collisions between such particles are not like collisions between speeding automobiles. When automobiles collide, destruction ensues. But when particles collide, creation does. Through Einstein's $E = MC^2$, the energy of collision is changed into mass: the mass of other particles, of lots of other particles—even of particles no one has ever seen before. Perl would search for his yet-heavier electrons in the spray of particles created in collisions at SLAC.

The drive from Stanford University to the accelerator runs past Nieman-Marcus, Lord and Taylor, and a golf course. Turning into SLAC, the ambience changes. Immediately, one is confronted by a guardhouse. Although it carries no sign instructing drivers to pull over and identify themselves, the checkpoint exerts a certain intimidating effect. I was never able to pass it without a faint sense that I was intruding where I should not. The place has that sort of official air about it.

In the lunchroom at SLAC, a blackboard has been installed on which people jot down equations, or sketches for new pieces of equipment. Throughout the complex, the floors are linoleum, the desks gray metal. The long, featureless corridors are hung with color photographs—not of the Matterhorn or the Golden Gate Bridge at sunset, but of pieces of scientific equipment. Just down the corridor from Perl's office is a lounge where members of his group gather to relax: it is airless, cramped, and has no windows. Down on the university campus, the architecture says "You are privileged to join us here in our elegant, tasteful, and refined contemplation of eternal truths." At SLAC the architecture says "Let's get going: there's work to be done."

But people at SLAC do not think the architecture is important. It is the accelerator that is important. A security checkpoint—a real one, this time—bars access to it.

Passing the checkpoint, one crosses from the high-density, suburban world of San Francisco Bay into what at first glance appears to be a wilderness: the lovely wilderness of rolling California hills; brown in summer, green in winter, dotted with live oaks and eucalyptus. Cutting in a perfectly straight line across the hills is a 2-mile-long metal building.

Inside, the noise level—an incessant, electrical, vibrating buzz—is so great that it is hard to carry on a conversation. The racket comes from a series of klystron tubes, receding seemingly without end into the distance in either direction. They look like the transformers one sees on telephone poles: their function is to feed electrical power into the accelerator proper.

This is out of sight. It is buried 20 feet underground: a metal tube, several inches in diameter, evacuated to a vacuum to permit the passage of the accelerated particles. As for these particles, they are of two types: on the one hand electrons, and on the other positrons, particles possessing the opposite electrical charge.

Within the tube is an intense electric field created by the klystrons. The field exerts a force on the electrons and positrons. Responding, they pick up speed as they travel down the tube. The electric field does, too. It hurries along behind. In form, it is a wave, not unlike the water waves down which bathers surf, and in just the same fashion the electrons and positrons surf down the traveling wave of electrical force—surf for 2 miles, at great acceleration.

The electrons and positrons travel in bunches. Each is about a millimeter long, far smaller in diameter, and contains perhaps 10 billion particles. If you could see one, a bunch might look like a tiny snip of human hair hurtling furiously down the tube. Such bunches of rapidly moving particles are prone to violent instabilities. They tend to fly apart—to dissipate. This is because electrons are negatively charged, and like charges repel one another. Similarly, the positrons are positively charged, and repel one another. So the bunches must be constrained. They must be held together. This is accomplished by a series of magnets placed at strategic intervals along the accelerator, which exert guiding forces upon them.

The particles zip along. Delicately herded together by the magnets, continually hounded by the traveling electrical field, they zip at nearly the velocity of light. They fly past the klystrons, past the live oaks and eucalyptus trees, underneath a freeway that crosses at one point overhead. Alongside the accelerator is a 2-mile jogging path. The particles make it down their 2-mile path in slightly over a hundred-thousandth of a second.

Finally, at the business end of the accelerator, gigantic bending magnets separate the electrons and the positrons into two beams. Perl's experiment required them to be fed into a donut-shaped storage ring, the electrons traveling one way and the positrons the other within it. The ring's name is SPEAR, which stands for Stanford Positron-Electron Accelerator Ring (is it surprising that the people of SLAC, which is pronounced like "slack," chose such a name?). At a precisely selected point in SPEAR, the beams are brought together. Brought together to slam into one another. And there all the millions of watts, all the millions of dollars, all the technology and giganticism and delicate care come focused to a point and the particles violently collide . . . collide within a detector, itself a complex, delicate instrument—an instrument that records the debris from that collision.

Perl had taken part in the building of that detector. It had been an extraordinary device, far surpassing the bubble chambers Alvarez had pioneered, and the task of building it had required the efforts of twenty people over four years. When he was young, good work in high-energy physics had been done by a mere few people working together. But by the early 1970s groups of twenty were standard. And nowadays the field has grown so complex, the machines so gigantic, that it requires several hundred scientists, working together over a period of many years, to get a result. (Luis Alvarez and his bubble chambers had been instrumental in moving the field along this path.) Lounging around Perl's office one day, I caught sight of a recent scientific paper: nearly half its length was taken up with the names of the scientists who had authored it.

Throughout the past century, science in America has undergone a steady growth. In the decade following the Civil War, no more than seventy-five Americans called themselves physicists. By the time of the Depression that number had swelled to twenty-five hundred. By the mid-1950s there were twenty-one thousand. Historian Daniel Kevles has recounted that at a meeting of the Physical Society one day, a member of the Old Guard "looked out over the sea of faces. Isn't it terrible, he lamented to a friend, how we don't know any of these new men? (Yes, the friend replied. But what's worse, they don't know us.)"

By now this growth has led to a fundamental shift in the very nature of science. But the growth has not been continuous. More than anything else, it was the destruction of Hiroshima that created big science. In an extraordinary passage, historian of science Gerald Holton has written of how

> a secret army of scientists, quartered in secret cities, was suddenly revealed to have found a way of reproducing at will the Biblical destruction of cities and of anticipating the apocalyptic end of man that has always haunted his thoughts. That one August day in 1945 changed the imagination of mankind as a whole—and with it, as one of the by-products, the amount of support of scientific work. . . . The traumatic experience of one brief, cataclysmic event on a given day can reverberate in the spirit for as long as the individual exists, perhaps as long as the race exists. Hiroshima, the flight of Sputnik and of Gagarin—these were such mythopoeic events. Every child will know hereafter that "science" prepared these happenings. This knowledge is now embedded in dreams no less than in waking thoughts; and just as a society cannot do what its members do not dream of, it cannot cease doing that which is part of its dreams.

Only since the Second World War has government felt itself obliged, as a matter of course, to fund scientific research lavishly. It is this transformation that has made big science possible.

And big science differs from small science, not just quantitatively, but qualitatively as well. The very *feel* of the profession, the nature of the workplace, has been dramatically altered by the shift. For centuries the model for science was the solitary individual, struggling to comprehend the universe armed only with his unique, personal creative fires. But in the world of big science, the model is the team. Not long ago, a paper signed by 365 people was published in a journal of physics. Life in such a group differs fundamentally from that of the world of small science. People in groups are specialists with a vengeance: nobody in a big collaboration could hope to do everything. Typically, the group will split into subunits, each devoted to a tiny fragment of the whole task. A subgroup might consist of ten to fifty members—in itself large as these things go—and each will work to some degree separately from the others. One might be building a spark chamber, another might concentrate on the design of the detector's magnets, a third on writing some small part of the immense computer programs that will be used to analyze the data.

The task of welding all these disparate units together is a central problem of big science. How can such an immense body of people—these groups are bigger than the entire faculty at Amherst College, my home institution—work together as a smoothly functioning unit? They do so by meeting. A group meeting resembles in many ways a convention, with the difference that these conventions are devoted to a single task. Talks are presented in which subgroups describe the status of their work. Often several have been working on the same problem, each exploring a different avenue of approach. Based on all these progress reports, decisions are made on what to do next.

These decisions are typically made by the senior members of the group. A big group will have a number of middle-level managers who preside over all day-to-day decisions, and at the top a group leader or, as he is often termed, a "spokesman." This spokesman—to the best of my knowledge there has never been a woman leader of a truly big group—is in many ways analogous to a university president. He handles matters of overall policy, is the point man for obtaining funding, and wields great power.

Two anecdotes illustrate the enormous transformation that has been wrought by the emergence of big science. The first concerns a young physicist who wished to ask Albert Einstein a few questions. Einstein agreed to meet him at a certain time on a certain city street. Unfortunately, the youngster was delayed.

Arriving hours late, he apologized profusely. But Einstein waved the apologies away. "There was no difficulty," he explained. "I can do my work anywhere." Einstein had continued his research standing on the street with the traffic rumbling by.

The second anecdote is recounted in Gary Taubes's *Nobel Dreams*, a portrait of the Nobel Prize–winning physicist Carlo Rubbia, spokesman of a large group:

> A woman physicist [in Rubbia's group] had been waiting for a couple of weeks to steal just a few minutes of Rubbia's time and discuss what she considered a crucial and highly important piece of physics. Rubbia also thought it was important but he had been flying around the world, coming and going, and the physicist had just about given up hope.
>
> Finally, one morning she gets a call from Rubbia. She picks up the phone and Rubbia says "Okay, I have exactly twenty minutes to talk to you about your work." This is great, she thinks. She slams down the phone, runs full speed to Rubbia's office, making it there in about ten seconds, only to find that his door is locked. She turns to Rubbia's secretary and says "Carlo's door is locked?"
>
> "Yes," the secretary replies. "Carlo was calling from the airport."

Rubbia was calling from the airport because he was a professor at Harvard. But the group he ran—with an iron hand—was in Switzerland. He had been flying in from Cambridge *every week*. His students, Taubes reports, had dubbed him "the Alitalia professor." Most of the time, the junior members of his group were not even aware whether he was around or not.

Successful practitioners of big science nowadays need to be more than good scientists. They need to be able to work well with industry and with government agencies. They need to be expert at public relations and fundraising. They need to be good coalition builders, masters of arm-twisting judiciously mixed with charm and persuasion. I think it is fair to say that the image the public holds of the scientist—and this is the image that most scientists hold of themselves as well—lies closer to the first than the second of my two anecdotes. But the spokesman of a high-energy physics group behaves more like the CEO of a large corporation than like Einstein.

The junior members of a group—and these are Ph.D. scientists—often feel like mere cogs in a wheel. They have little say in any major decisions: often they have little idea of what is even going on beyond the confines of their own subgroup. "These young people simply have to take orders," Perl told me. "It can take three years, five years to build a detector nowadays—not doing any science, just constructing the instrument. And out of all the people in that group, how many will become known, famous? Only a few. And what is going to give all the rest of them their professional satisfaction? How are they going to obtain recognition?"

The question of recognition, of credit for fine work, goes to the very heart of how we think about the work of the scientist. All of us, scientists and nonscientists alike, associate great discoveries with individual people. We credit Darwin with the theory of evolution, Einstein with the theory of relativity. And as we admire Beethoven or Shakespeare, so we admire these people for what they did. But who will admire some anonymous member of a huge group?

Starting in the 1930s and continuing to this day, an important body of work has been published by a mathematician named Nicolas Bourbaki. Bourbaki must be a remarkable person, for his publications contain some very remarkable mathematics. Unfortunately, however, nobody has ever met him. This is because Bourbaki does not exist—he is a mythical individual, coming from the mythical university of Nancago, in the mythical country of Poldevia. In reality all "his" works are written by other people—by lots of other people, by a group of famous French mathematicians who have decided to gather together and publish their work anonymously.

What is the point of this story? The point is that it *is* a story, a story that every scientist and mathematician knows. Bourbaki is famous precisely because such voluntary anonymity is so rare. We all regard it as extraordinary that anyone would do such a thing. The general public might think the scientist works selflessly, seeking only the advance of knowledge. Not so: all of us are out there looking for glory. I think it is significant in this connection that the members of the Bourbaki group are all well known in their own right. Only those who have already attained recognition can afford the luxury of anonymity.

At present, problems of attribution are dealt with informally, through the "whisper network" whereby group leaders, or middle-level managers, quietly put out the word as to who should get the best jobs. But I would argue that in the long run this will not do. More and more fields are moving from small to big—my own field, astronomy, is a case in point. Furthermore, those already big are getting bigger fast. Very soon, methods will need to be found of formally establishing credit in big science, methods analogous to those that already exist in small science.

Not long ago, R. R. Wilson, one of the world's most noted practitioners of big science, published an autobiographical memoir. The title he chose was *My Fight Against Team Research*. It was only partially a joke.

In many ways big science is an evil, albeit a necessary one. There are many drawbacks to it. In the good old days, if a new idea for modifying your experiment occurred to you during breakfast, you would simply make the change when you got to the lab. Such flexibility is impossible now. No matter how good your idea, no matter how appropriate the change, it will have vast repercussions, and vast numbers of people will have to be consulted. If your scheme is accepted, documentation must be provided, setting forth all the details so that those who come after will understand what has happened. Small science is like riding a bicycle: big science is like piloting an ocean liner.

Some groups are so big that its members never even meet each other. The famous—or rather, infamous—collaboration among 365 physicists is a case in point. The members of this group come from thirty-three institutions in four nations: many of them met for the first time at a conference *after* finishing their project. Had the Super Collider become a reality, the problem would have been exacerbated: prior to its cancellation, more than 900 physicists had signed up to work on a single experiment being planned for the machine.

Big science is by nature conservative. You cannot afford to take very many risks if your experiment costs hundreds of millions of dollars. Even if you wished, your funding agency wouldn't let you do it—or at the very least, would refuse to renew your contract. Nor would any collaboration consisting of hundreds of people, each of whom has given years to a project, risk all on a long shot. Only individuals, working with small budgets, are free to do such things.

Notwithstanding his mastery of the field, Perl is disturbed by the trend toward ever bigger science. He worries that giant groups stifle creativity and new ideas. Bright, inventive young physicists, anxious for the freedom to work independently, are heading elsewhere, he fears. High-energy physics, in his estimation, is becoming afflicted with myopia: the more time passes, the more the field concerns itself with fewer and fewer issues. One of the currently popular avenues of research is the search for the so-called Higgs particle, predicted by theory but not yet found. "There are books on how to find the Higgs particle," Perl says. "Meetings on how to find the Higgs particle. The thing to do if you are young is not look for the Higgs particle. I would not spend my time looking . . . I would try to look for things other people consider to be trash."

Can big science ever produce great science? Can the brilliance, the quirky creativity required for great accomplishment ever flourish in such an environment? The old adage "no committee ever wrote a great novel" reminds us of the inherent leveling effect of large collaborations. Remarkably however, history teaches otherwise. Great discoveries have indeed come out of large collaborations. In a recent twenty-year interval, four Nobel Prizes went to work done by large teams.

I think it is significant, however, that none of these Nobel Prizes has gone to a group as a whole. Rather they have been awarded to the leaders, or to certain selected members, of the groups. But why? Is it really true that these individuals have succeeded in stamping the work of the entire team with their personal imprimatur? Have they controlled their groups so thoroughly as to render them little more than extensions of themselves? If Amnesty International can win a Nobel Peace Prize, if the New York Mets can win a World Series, why cannot the several hundred people who have collectively made a great discovery win a Nobel Prize in physics?

My guess is that this peculiar circumstance is no accident. I believe it stems from the image we hold in our minds of the scientist as a solitary individual, struggling to bring forth out of the depths of his or her soul the creative fires necessary to pursue great work. If I am right in this suspicion, then it is easy to see why people find it hard to believe that great science can be done by huge collaborations: our mythology defines this to be impossible.

But is it really true that no committee ever wrote a great novel? Or is this a romantic fiction? It seems to me that great works of art often emerge from huge collaborations. An opera production results from the combined efforts of large numbers of highly skilled individuals working together as a group. The same is true of movies. Every Gothic cathedral is eloquent testimony to the power of group work. The design of a big accelerator has all the brilliance, all the logic and sense of inevitability of a Bach fugue. A particle detector is every bit as intricately crafted as the Book of Kells. But these great instruments lie entirely beyond the capacities of a single individual. It was group work that wrought these wonders.

At the close of our tour of SLAC, Martin Perl brought me to the room where the data was gathered. Here was where all the planning, all the fundraising and group meetings and instrument design came to fruition, and research in high-energy physics was actually underway. Even as we watched, an experiment was in progress. Particles were being accelerated to enormous energies and smashed together. Half a dozen people sat before computer consoles. Before them on their screens were the starburst images of particles flying outward from the point of collision with unimaginable velocities. Peering over their shoulders, I realized I was looking straight into the heart of the cauldron. I was seeing the Central Fire. I was witnessing what happens when matter—substance, stuff—is strained to the very breaking point and beyond.

Martin Perl was suddenly seized with a shiver of excitement. A surge arose within him. "Do you see?" he asked. "Do you see now what makes this work worth it after all? The six people sitting here have spent years of their lives building this detector. They are members of a collaboration consisting of several hundred people. But right now, it is they and they alone who are running the experiment. At this moment, these six people are in total control of one of the finest scientific instruments in the world."

In the end, it had come down to this. A gigantic inverted pyramid stood balanced upon its point. SLAC and the detector they had built added up to one of the few instruments in the world capable of carrying on research in high-energy physics—and these few people were dominating it. They lounged around carelessly, sipping coffee, glancing at their screens. From time to time, someone leaned forward to make a minute adjustment. The data were streaming in.

Perl and his group had gambled back in the early 1970s that their new particle, hypothetical third member of the sequence beginning with the electron and the muon, might exist. If so, it could be found in the debris produced in a particle collision. But found how?

The difficulty was that it would be invisible to their detector. It would be invisible because it was expected to decay—to fragment into other particles—long before any detector could record its existence. Based on what they knew of the electron and the muon, they expected their particle to survive a mere ten-trillionth of a second before decaying. But even traveling at its great speed, it would have moved no more than a few hundredths of a millimeter from its point of creation before vanishing. No detector imaginable could probe events with so fine a resolution.

They would have to search, not for the particle, but for what it split into: its decay products. What would these products be? The muon was known to decay into an electron plus other particles that could not be spotted by detectors. By extension, the new particle would be expected to do this, too. Alternatively, it could also decay into a muon (plus the undetectables). Since the new particles would be produced in pairs, their decays could yield a pair of electrons, a pair of muons, or one of each.

The first two of these possibilities were no help—pairs of electrons and muons could be made in lots of other ways. But an electron accompanied by a muon would be interesting. Perl decided that these events would be his signal that the new, third member of the electron family had been created at SLAC. His experiment was designed to search for such electron/muon pairs. The experiment began running in 1973: within a year, the pairs began showing up in the data.

One might imagine a great uncorking of champagne bottles at this point. Reality, unfortunately, was otherwise. The scientist does not simply "look" at the natural world, gathering facts the way we might pick up shells upon a beach. Data is not so easy to get hold of as that. In truth, what we call scientific facts are thickly encrusted with layer upon layer of interpretation. Rather than having discovered the particle they were looking for, Perl and his group might merely have made a mistake.

One possible source of error concerned the so-called D particle, itself recently discovered. Like the object of their search, the D was capable of decaying into either an electron or a muon. The difference was that these decays also involved emission of photons, or particles of light. Thus the electron/muon events, if caused by the D, would also be accompanied by photons. A clean test: they searched for the photons, hoping never to find any.

The task sounds easy. Unfortunately, it was not. Their detector was imperfect. It had blind spots, nooks and crannies through which the light might pass undetected. Perl and his group were forced to study these imperfections. Eventually, they decided that roughly half of any emitted photons would go undetected. Were, then, the electron/muon events accompanied by photons half of the time? Again, while on paper the test was simple, in reality it was not. For their detector did not just have blind spots. Like all scientific instruments, it was also plagued by errors, random fluctuations in its behavior. Operating at the limit of sensitivity, at the cutting edge of the possible, these random errors loomed large. Furthermore, they were inescapable—Perl did not *expect* perfect data. Had he found any, he would not have believed it. Rather, he expected some fraction of the measured characteristics of each event to be just plain wrong. And as it is possible to toss a fair coin ten times and get nothing but heads, so he expected to be seriously misled on occasion.

They sweated it out. Another problem was that the D particle was popular. High-energy physicists loved the D—for two reasons: it was a new thing, recently found, and it was the first member of an entirely new class of particles, long predicted but only now discovered. In contrast, there was not the slightest theoretical reason to expect Perl's "next electron" to exist. Who needed such a thing? People went around saying that he was being fooled by the D. It did not help that the D was SLAC's baby, having been discovered there—indeed, Perl had been a member of the group that found it.

There were other mistakes they might be making. The muon, unlike all other particles, was capable of penetrating great quantities of shielding. Indeed, this was how they identified it: anything that made it all the way through the detector was simply declared to be a muon. But was it? Another possibility was that some other particle was being produced, traveled part-way through the detector—and then decayed into a muon. Here, too, they formulated a test: only about one out of five such interlopers was expected to fool them in this fashion. But here, too, imperfections and random errors plagued the test.

Meanwhile, there was competition. A group in Hamburg was conducting a similar experiment. Not only that, but they were ominously *not* finding electron/muon events. But why not? If Perl was right, the Germans should have found many. He roamed the corridors restlessly at SLAC, wandering into people's offices to think the matter through out loud. He invented scenarios, trying to imagine what mistake the competing group might have made. Or that he might have made.

One evening while visiting Perl, I accompanied him and several of his closest friends to dinner. It was a delightful evening, filled with good food and laughter. In the middle of the meal, and entirely without warning, Perl commented how strange it felt to be having such a good time when so many people in the world were suffering.

There is a gloomy element to him. Maybe it comes from his childhood during the Depression, when his father, an independent small businessman, was continually threatened with disaster. This moodiness pervades his scientific work as well. "A very important part of the way I do things," says Perl, "is a high level of anxiety. I am not calm about these things. I stew over them. The job has to be pushed through, and it has to be done right. Many were the times during that project that I wished I could abandon the whole damned thing. During the bad stretches, I was nervous and anxious. I was losing a lot of sleep."

It was a solitary time. Perl knew full well that he was far out on a limb. Even members of his own group doubted they had discovered anything important. But throughout it all, he never wavered in his conviction that they really had found the "next electron." "It was such a natural explanation for the data," he says. "It just worked. You didn't have to twist and turn, you didn't have to fiddle. All the other possible explanations for these events somehow involved contrivances. But the idea that it was our particle producing these electron/muon events [in our detectors] just cleanly reproduced the data. Another issue was that the signal was always there. We would run at different energies, we would change some element of the experiment—and the electron/muon events kept on appearing. They were robust."

Bit by bit, Perl and his people worked their way through all the possible sources of error. Eventually, the other members of his group came to agree with him—and then the rest of the world of high-energy physics, too. Ultimately, a different group working at the German accelerator began finding electron/muon events as well. To this day, Perl is not sure why the first had failed to find them.

By 1977 they published their result. They had begun their search four years earlier, in 1973. Throughout all that time, there was no particular moment at which one might say the new particle had been discovered. Rather, the story is one of a slow evolution over a period of years, in which it gradually emerged from limbo. During all their work they had contented themselves with calling their hypothetical particle U—for "unknown." But eventually, once its existence was assured, the time came for a symbolic act: its formal naming. After much hemming and hawing, they settled on "tau," first letter of the Greek word for "third." Perhaps the fact that a graduate student in the group was Greek had something to do with the choice. They had also asked the group secretary: she found the letter easier to type than the other candidates.

At lunch one day in the cafeteria at SLAC, Martin Perl and I were joined by two young theoretical high-energy physicists. Lunchtime conversations among scientists are often a rather competitive game, the object of the game being (a) to know more than everyone else, (b) to say what you know the fastest, and (c) to say it the most bitingly. In my experience, high-energy physicists are more competitive players than workers in other fields, and theoreticians more competitive than experimentalists. So the conversation that day moved along at the pace of a hard-fought professional tennis match. But throughout it all, Perl remained silent, smiling benignantly and munching his lunch meditatively. An outsider coming upon the scene might have mistaken him for a kindly uncle, over for a visit and hopelessly out of his depth.

Is Martin Perl shy? No one would have thought so watching him chase down the tau. This was the work that won him the Nobel Prize. The other members of his group had little faith in the existence of the particle. The search was his baby from start to finish. It was he who pushed the project; it was he who toured the country, arguing for the tau's existence; it was he who insisted they publish their findings, while other members of his group were still paralyzed with the fear they would be making fools of themselves by going pubic. He even held a press conference, announcing their discovery. At every stage of the process, Perl wanted there to be none of those doubts that so often crop up, years after a great discovery, as to who had really made it. He was aggressively claiming turf.

But Perl exhibits a fascinating combination of aggressiveness and shyness. He is tall, loose-limbed, with graying frizzy hair and a gentle demeanor. He moves slowly and speaks slowly — there is something almost courtly about his manner. He dislikes crowds, hates standing around at parties with drink in hand, and at conferences will often eat dinner alone in his hotel room. Indeed, even in the company of close friends, he is reserved in groups. But this same man is perfectly at home managing a big research team, holding a press conference, or delivering a lecture to a crowd of thousands.

At one point during our tour of SLAC, Perl fell into a conversation with a somewhat garrulous security guard. I saw no trace of condescension in Perl, nor of restlessness as the conversation showed few signs of winding down. But there are elements of the slave driver in him. Until relatively recently, he would feel not the slightest qualms about rousing junior members of his group out of bed at 3:00 A.M. to ask a question during an all-night run. On one occasion he phoned a technician to shout at him—literally—about the quality of his work. Things got so bad that at one point the SLAC director called Perl in and told him to calm down.

Perl's reaction to these episodes is revealing. He is ashamed of them—not because they were inappropriate, not because people should not do such things, but because those on the receiving end of his imperiousness had no recourse. "I have to be careful about these things," he told me, "because people under me are so trapped. I am the group leader, and these people are my underlings. They are powerless."

Perl has a strong sense of social justice. He is sympathetic to the labor movement and he supports unionization at SLAC. His politics are very left, and he is deeply distressed by the fragmentation of the left in recent years—the anti-Semitism of some black leaders troubles him. Not long ago, he joined an effort to press the Stanford administration to provide the same benefits for same-sex and unmarried couples as for traditional ones.

Perl is Jewish and, although not religious, is a Zionist. Born in 1927, he was a communist sympathizer for many years. For a while, after the Second World War, he took night classes on communist thought. In 1948 he marched in the May Day parade along with the Abraham Lincoln Brigade—"a thrilling moment for me," he said. In those days he accepted Stalin's purges "because the victims were left-deviationist, and had to be punished. Even now, I am sad the Soviet Union has collapsed. Communism posed a possible alternative to capitalism, and it is too bad the alternative turned out not to work. In reality, of course, the Soviet experiment turned out to set back the cause of socialism. But it had not been until the 1950s that I realized that Marx had been wrong, and that capitalism was not collapsing of its own weight. Looking back, I realize now that I had been very naive in those early days. Not until the early 1960s did I let myself realize what the Stalinist purges had really meant."

During the same years in which he was chasing down the tau particle, Perl was also deeply involved in social issues. He spoke out against the Vietnam War, and against scientists' work in support of that war. He organized a conference on the job market for young physicists. He wrote on the failures of the system whereby the government seeks scientists' advice on policy matters, only to ignore that advice. He wrote on the social responsibilities of scientists and professional scientific organizations. He was one of a small and influential group that founded a number of organizations dealing with issues concerning the relationship between science and society. As a direct result of his activities, a committee of the American Physical Society now undertakes detailed technical studies of government activities, and takes active political stands based upon its conclusions.

In July of 1983, a panel of prestigious high-energy physicists submitted an ambitious recommendation to the Department of Energy: the construction of an accelerator to end all accelerators—a behemoth, a giant: the Superconducting Super Collider (SSC).

The panel's proposal was prodigious in its scope. One component of the Collider was to have been a device that reached energies far greater than SLAC's—but this was to have served merely as a sort of injector, a preliminary stage of acceleration before the monster got down to seriously pouring energy into particles. The Superconducting Super Collider would have been bigger than a city.

The SSC would not have been big science. It would have been megascience. Whipping protons about in a circle 52 miles in circumference, through a tunnel the length of the Washington Beltway, the Collider would have reached an energy twenty times higher than that provided by any other accelerator in the world: this is the comparison between the speed of a jetliner and that of a bicycle. Thousands of superconducting magnets, each refrigerated to a temperature of $-452°F$, would have guided the particles in their course. The behemoth would have consumed 2,000 gallons of water per minute, and 250 million watts of power. It would have required the services of twenty-five hundred maintenance workers and technicians. During construction, towns nearby would have had to accommodate up to ten thousand additional residents. When completed, the Collider was to have a staff of one thousand permanent and visiting scientists. For comparison, there are only about twice that number of high-energy physicists throughout the entire United States today.

Everything about the project was colossal. The mere task of *writing* the formal design proposal cost nearly a million dollars. That proposal ran to four hundred pages and represented the efforts of 150 scientists. (They met at Berkeley: "To show you how seriously Berkeley takes all this," a member of the group wrote, "they have given all of us parking spaces.")

The report of that design study group nicely combined the gigantic with the bucolic.

Resembling some vast Stonehenge on the open landscape, every two to four miles apart is a small cluster of structures housing power, refrigeration and other facilities, connecting to a tunnel hidden some twenty feet below ground level. At six other locations around the huge ring are larger complexes of buildings identified as collision halls filled with detectors for experiments. Where the ring passes beneath farms, crops grow and cattle graze. Where its course takes it under forests, rivers and hills, nothing appears to be disturbed. Where it lies below highways, suburban housing or shopping malls, life goes on as usual.

President Reagan announced his support for the SSC early in 1987. Money was in the air: lobbyists for the concrete industry began referring to the project as "the big pour." The Department of Energy announced guidelines for competitive bidding as to the site selection. Forty-three tons of proposals from twenty-five states landed at their door. Several could be rejected outright—one called for placing the thing in orbit about the Earth—but the task of selection led to numerous headaches. When a site in Texas was ultimately chosen, several loser states threatened lawsuits: one congressman asked the General Accounting Office to determine whether the selection process had been fair.

Cost projections started big and got bigger fast. An early estimate in 1984 stood at close to $3 billion. Within two years, the projected cost had inflated to $4.4 billion: by 1987 it was closer to $5 billion and, by 1990, $7.8 billion. By the time Congress ultimately killed the project in 1993, one projected estimate was a staggering $13 billion. Projections of the time required to construct the accelerator similarly inflated. Scientists working on the machine lived in a perpetual twilight state of anxiety. Visions filled them of the cost rising to infinity, and the date of completion receding into the next geological era.

From the very start, the SSC project aroused intense controversy—even among physicists. Recently, I spent a few days going through a decade's worth of letters about it to the editor of an influential physics journal. They made for interesting reading. One element that struck me was the degree of acrimony evinced, often on both sides of the question. Lots of people were angry. A second point that struck my attention was that the debate never seemed to be getting anywhere. The issues raised during the first few years just kept on resurfacing, year after year as the debate rolled by.

A quick summary of the issues raised in these letters might go as follows. (1) With all those poor people out there, how can you justify spending so much money on pure science? (2) But if you took that argument seriously, you would never support *any* basic research. (3) Why high-energy physics: why not my field? (4) Because high-energy physics is so fundamental. (5) No, it isn't. (6) The SSC will generate lots of technological spinoff. (7) No, it won't. (8) The cost is too great. (9) The cost isn't so great compared to the space program—not to mention defense.

Some of these issues are valid. Federal support for high-energy physics is indeed lavish compared to that for any other field— roughly fifty cents out of every federal dollar for physics goes to high-energy physics. And no matter whose estimate you believed, the SSC would have been stupendously expensive. SLAC, for instance, had cost a mere $114 million in 1966 dollars to construct. That figure translates into about half a billion 1993 dollars, so for the cost of the SSC something like twenty new SLACs could have been built. Was such an immense outlay worth it?

One other argument perpetually raised in this debate needs to be addressed. It is that the SSC's cost, if spread around evenly to increase funding for *all* basic research, would have a far greater benefit. I would argue that this is perfectly true: science is intrinsically anarchistic, and it works best when large numbers of investigators independently pursue their own projects. But I would also argue that the fact is irrelevant. Congress never would have appropriated such a sum for science in general.

George Keyworth, President Reagan's science adviser, once said in a speech, "I won't conceal my opinion that it would be a serious blow to U.S. scientific leadership" if the SSC were not built. The machine would be "a magnet for talent and creativity . . . involving far, far more than the relatively small number of people who can work directly with it, because it stimulates interest in science and excellence far across society and because it inevitably spins off new ideas." I accept Keyworth's argument. Only a single project like the Super Collider, with all its attendant glory, is capable of inspiring the public whose taxes pay for it. Big science thrills in a way that small science does not. Imagine the excitement, particularly to the young, of visiting an entire city devoted to knowledge.

Sheldon Glashow and Leon Lederman, both staunch advocates of the SSC, sounded the call to action:

The universe astonishes us by its very comprehensibility. In this we find our call. Being born upon an obscure planet located at the rim of a middling galaxy among a hundred billion galaxies of an aging universe, it is our sacred duty to know its deepest secrets, as well as we are able. Dolphins and chimpanzees can be made to speak, after a fashion. Yet, only humans will look at the stars with wonder and find it necessary to understand just what they are and how they work and why we are here to see them. No better mouse trap or wrist TV here—just the triumph of human imagination.

The passage positively bristles with quasireligious awe, and it is badly overdone. But for all of that, its sentiments ring true. Yes, the Collider was expensive—but just how expensive? Thirteen billion dollars, spread evenly over the entire population of the United States, adds up to barely more than $50 per person. Furthermore, this cost would have been spread out over the many years of construction. Should Americans really be so unwilling to spend several dollars a year to learn a little more about the fundamental nature of matter?

Glashow has written,

Who knows what surprises the SSC will reveal? If we did we wouldn't need the machine. We don't quite know what we are doing nor where it will lead. That's what I mean by fundamental, and it's really the only honest argument we've got going for us.

While many physicists opposed the SSC, most of those who would benefit directly from it—the high-energy physicists—actively supported the project. Martin Perl was one of the few who did not. Even after billions of dollars had been spent on the machine, he advocated closing the project down. His hunch had always been that we would not need to go to such enormous energies in order to find new and exciting things. He feared that the effort required to get the machine built would be so immense that the project would leave high-energy physics exhausted and enervated. And he feared that the claims made for the SSC might not come to pass. "What are we going to say to the American public if the machine turns out not to make all those wonderful discoveries?" he asked. "Scientists have the tendency to promise taxpayers the Moon when pushing for their pet projects—and then forgetting the American public once funding is assured. We can't keep on doing this forever."

The reader will have noticed that I was sympathetic to the SSC. Nevertheless, one element of the project always disturbed me deeply. It is the machine's contribution to the steady centralization of high-energy physics. In 1953 there were twenty-two accelerators in the United States. By 1965 that number had shrunk to eight, and by 1980 to four. Had the SSC been built, there would have been precisely one location at which frontline research on the nature of matter was being conducted—and not just in the United States, but throughout the world. But such centralization frightens me. The push to ever bigger scientific projects is simultaneously the push to ever more centralized control of science. I am convinced that a certain anarchy is good for the field. Science flourishes best when people are independently pursuing many different avenues—many different kinds of avenues.

In the end, of course, all has come to naught. The SSC began with a bang but quickly settled into a prolonged whimper. In hard times, with numerous federal programs being slashed, the accelerator was an obvious target. The dissension within the ranks of physicists did not help, nor the gigantic escalation in cost. Prior to site selection, plenty of congressmen supported the machine; but with the choice of Texas, much support on the Hill dried up overnight. It did not help the SSC's chances with Congress that the Speaker of the House, Texas representative Jim Wright, resigned not long after the site selection; and that Texas senator Lloyd Bentsen, chairman of the Senate Finance Committee, soon moved over to the Department of Treasury.

For the last several years of its life, the project teetered perpetually on the brink of doom. Charges of mismanagement were rampant: the Department of Energy promised a drastic shakeup. A veteran of the Naval Sea Systems Command and the Naval Ships Weapons Engineering Station was appointed project manager. The rumor was that relations were so bad between

him and the scientific director that by the end they were com-
municating solely by memo. At the construction site, people
lived in a world of endless audits and inspections and governmental oversight.

In 1991 the Collider received funding only after an acrimo-
nious debate in the House. The following year the House voted
by a wide margin to kill it: the Senate voted for it, and the project was saved in conference. Then, in the fall of 1993 the House
voted to kill the SSC yet again, this time by a yet wider margin.
Again, the Senate voted for it. But this time, in conference, the
House members were adamant. Bowing to the inevitable, the
senators in the conference committee concurred in the vote to
terminate the SSC.

The fate of the Super Collider points to one of the gravest problems of big science: its dependence on the vagaries of politics.
Big science and small science differ profoundly in this regard.
Small science depends for its support on various federal agencies, such as the National Science Foundation and the National
Institutes of Health. These agencies operate in a manner designed to provide the most rational support for the progress of
science. One of their chief functions is to formulate a set of policies intended to encourage integrated development of an entire
discipline. The National Science Foundation, for instance,
might decide to decrease its support for planetary astronomy,
recognizing that NASA has recently increased its funding level,
and choose instead to increase support for solid state physics.
The National Institutes of Health might decide that throwing
more and more money at cancer has yielded little benefit in recent years, while a relatively small sum spent on diabetes might
make a big difference.

Such policy decisions are at least partially insulated from direct governmental influence. And at the level of the individual
project, the insulation is even greater. Every request for support
received by these agencies is sent to an outside scientist in the
same field for evaluation. Thus the choices of which particular
projects to support depend primarily on the judgments of other

scientists, not of government officials. Small science is run by professionals.

But big science is run by amateurs—by politicians. Every big science project is funded, not by a federal agency, but directly by Congress. Funding for the SSC was tacked on as an amendment to an Energy and Water Development Appropriations bill, of all things. Congressional support for these projects, in turn, often rests on concerns unrelated to the scientific merit of the proposal. Those projects get funded that scream the loudest— that attract somehow the broadest public support. It was not accidental that many congressmen dropped their support for the SSC once they learned that the lucrative construction project had been awarded to another state.

Furthermore, the requests that go to Congress are not part of any coherent overall plan for the development of science. Congress is regularly besieged by scientists pressing for their various projects—biologists pushing for the human genome project, astronomers for the Hubble Space Telescope, high-energy physicists for the SSC. These scientists have made not the slightest effort to coordinate with one another. Indeed, they are usually in competition.

I would argue that this is no way to do things. A congressional committee is hardly the place in which to formulate a rational approach to science policy. Congressmen are not qualified to do this—they don't *want* to do it. As more and more fields of science become big, as big science grows yet bigger, it grows ever more necessary to fund huge scientific projects in exactly the same way that small ones are: by a coherent process of planning. Among all the fields of science, I know of only one—my own field, astronomy—that regularly formulates an overall plan of development. Each decade, a committee is chosen and charged to survey the field in its entirety, and to formulate a set of recommendations for new projects, ranked in order of priority. That is the best possible way to approach the federal government.

For let no one think that anything positive has come out of the SSC debacle. As things now stand, it would have been far better had the project never been started. Two billion dollars have already been spent on the machine: that is enough to build four new SLAC's. Ten miles of tunnel now lie vacant beneath the Texas soil. The design studies that went into the SSC have little relevance to any other machine. The technology that has been developed for it may prove to have some residual spinoff, but surely not enough to justify the effort. Nor will that empty tunnel in Texas prove useful to many. For most practical purposes, the SSC has come to nothing.

Who paid for the SSC? Who pays for all of science—for all those wonderful instruments upon which science depends? Certainly not the people who use them. I do not know a single scientist who contributes a penny of his or her own money toward the costs of his labors. It is the taxpayer who supports science.

But science is elitist. No scientist would tolerate for a moment any interference by the taxpayer in his or her day-to-day work. I include myself in this blanket generalization. So far as I am concerned, only an expert has a right to an opinion on matters technical. I would never presume to tell an airplane pilot how to do his or her job. What do I know about flying a jumbo jetliner?—were I to get my hands on the thing, I would surely crash it. And by the same token, I would not listen for an instant to the opinions of a layman as to the best way for me to do my work. A sure recipe for disaster would be to allow the general public to direct the progress of science.

The taxpayer is therefore placed in a difficult position: asked on the one hand for money, and told on the other hand to keep his or her mouth shut about how this money is to be spent. I must say that there are times when my colleagues could be more tactful in their requests for funding. Sometimes their requests sound suspiciously like demands. I have heard scientists speak as if the public *ought* to support their research, as if science were some high calling that society was morally obligated to fund. The very phrase "pure research" conveys an exalted sense of holiness—an assumption of superiority, and the conviction that what the rest of poor, benighted humanity does with its time is somehow low and impure.

The collision between scientists and the Congress over the SSC is part of a larger issue: the tension between scientists and the society that supports them in their work. Congress is the flashpoint at which this tension is most acutely felt. A revealing interchange once occurred in a congressional committee hearing in which a scientist was pleading for yet more funds. "There is no end to scientific ambitions to explore," Congressman Chet Hollifield snapped. "But there is an end to the public purse. Every scientist that comes before us thinks his crow is the blackest."

This tension is inevitable. I do not think it will ever go away: by its very nature scientific research is expensive, and by its very nature scientific research is elitist. I'll go further: the tension is sure to grow as time passes. The SSC was only the beginning.

I believe that the natural world is infinite. We will never fully plumb its depths. No matter how far science progresses, there will always be questions, always mysteries whose solution we will seek. But I do *not* believe that the Stanford Linear Accelerator is infinite—nor any other scientific instrument. These are limited. Eventually, the time will come when they have outlived their usefulness. Eventually, the time will come when we will have to go beyond them. If we do not, science will stagnate. But that will take more money.

Gerald Holton has given a fascinating analysis of how a scientific field develops over time. Whenever a new machine is built, whenever a new theory is invented, a vast new continent is suddenly opened to exploration. Unexpected vistas beckon. In these early stages, discoveries come easy—they lie around like nuggets of gold strewn upon the grass. With a relatively small expenditure of effort, great strides are made. Attracted by the easy riches, new people—often the brightest people, often the younger ones—rush into the field.

But as time passes, all the easy work gets done. Now it is harder to make a discovery. The few remaining nuggets of gold lie hidden beneath a bush or buried underground. Now we must search to find them—search at great expense. (A former director of SLAC has noted that each great discovery in elementary particle physics has cost more than those that came before.) The smartest workers start looking around for some other fresh field to conquer. Ultimately, the once glorious new machine has been reduced to the status of an unimaginative drudge, devoted merely to the dotting of *i*s and the crossing of *t*s. All the exciting action is elsewhere.

Holton argues that this is the fate of *every* field, of every new machine. Just as each person grows old and dies, so does the progress of science inevitably tend toward stultification and decay. There is only one way for science to remain vibrant. It must perpetually renew itself: perpetually discover new theories, new points of view—and perpetually build new machines. Even had the SSC been built, it would not have been the end. It would not have been the last accelerator ever to be constructed. Ultimately, we would have had to pass beyond even it.

But will the public pay for all this? Or will science eventually grow too expensive for the public to support? This is no idle question, for I am painfully aware of one vibrant and exciting field of scientific discovery whose development has been essentially terminated for lack of funds. It is the study of the geology of the Moon. The Apollo missions to the Moon brought back a treasure trove of lunar samples. But the enormous expense of these missions has precluded further work in this area: not since 1972 has anyone set foot upon the Moon. The field has lost its zip.

We are accustomed to thinking that every problem can be solved, that every obstacle can be surmounted. But perhaps this is merely a fond wish. Perhaps everything has its limits. What is the biggest elementary particle accelerator that can ever be built? Surely, we will never build one bigger than the Earth. Surely, no nation will build one bigger than the size of that nation. Surely, no nation will build one more expensive than some significant fraction of its gross national product. Perhaps in the distant future a machine like the Super Collider will be built. No matter: the experience of the SSC reinforces the truth that at *some* stage the process is sure to terminate. And when this happens, elementary particle physics as a strong and vibrant field of study will come to an end.

Or will it? Perhaps there is another possibility. It is suggested by the spectacular progress of computer technology. The astonishing thing about computers is that they have grown smaller, not bigger, as time has passed. And at the same time, they have grown more powerful. While the advance of computer technology has led on the one hand to such behemoths as the supercomputer, it has also led to the PC. The sort of machine that eighteen-year-olds lug off to college nowadays dwarfs utterly the expensive giants of a mere few decades ago. Nor does there seem to be any end in sight to this progression.

Could such a thing happen to elementary particle physics? To study elementary particles, you need high energy—but to reach high energy, does your accelerator need to be gigantic? Perl wonders if perhaps it does not. "I am struck by the fact that most of the energy consumed by SLAC does not end up in the particles it accelerates," he says. "Rather, the bulk is consumed by the guiding magnets. While the energy of each particle emerging from SLAC is huge, there are so few of them that the total is actually rather modest.

"I take this to imply that no fundamental law of physics forbids us from building a *small* accelerator capable of reaching huge energies. We just haven't learned how to do it yet. SLAC, and every other accelerator in the world, operates using nineteenth-century physics. But maybe some day a smart person will figure out how to do better using the physics of this century.

"How could we do this? I have only the vaguest of notions. Particles are accelerated by an electric field, and the electric field produced by a laser dwarfs that produced by any modern accelerator. Alternatively, certain crystals have huge internal electric fields. Could either of these be used to accelerate particles? I have not the slightest idea—but perhaps it's not impossible."

If such a miniaturization were to occur, accelerators capable of reaching great energies would plummet in cost. Rather than requiring the SSC to do frontier research, you might only need something that would fit on your desk. The result would be an extraordinary transformation of high-energy physics. Suddenly, the giant research group would be a thing of the past. Frontline research on elementary particles could be done by individuals—by lots of individuals, scattered throughout universities and colleges and high-tech industries, each working independently to chase down his or her own quirky notions. Suddenly, the exhausting and chancy business of obtaining funding directly from Congress would cease, and people could go back to the far easier task of approaching the traditional agencies for support. And suddenly, the trend toward centralization so characteristic of big science would reverse. The healthy anarchy of small science would return.

And as for the mighty instruments of high-energy physics that we have today—they would be shut down. The throbbing, buzzing shed that is SLAC would fall silent. The whole length of the great accelerator would be carved up for its component parts, or boarded over and left to molder. Tour groups of the future might pass through the place, the guides mumbling by rote their litanies. Curious passengers would stare out through bus windows, gazing at the accelerator as we look upon the great pyramids of Egypt—ancient, outmoded monuments of a vanished age.

If such a transformation does *not* occur—what is the fate of high-energy physics? Will it choke itself to death out of greater and yet greater giganticism? And more than that: is *every* field of science fated to die in such a way? Is it possible that science itself will prove in the long run to have been but a passing phase? Have the centuries since the scientific revolution been but a brief flowering—a temporary phenomenon, a golden age, never to be repeated, in which we found it possible to learn, at not too great an effort, something more about the grand design?

At the close of our tour of SLAC, Martin Perl brought me to SPEAR, the storage ring in whose collisions he had discovered the tau particle. For some time now, the ring had been replaced by larger devices. The place was silent and deserted. A faint, dusty odor seemed to have settled over the place.

His discovery of the tau, third particle in the sequence commencing with the electron and muon, has sparked a spate of searches for yet a fourth particle. None have been successful. The fourth particle, if it exists, has never been found.

In one sense, Perl's original hope in searching for the tau has not come to pass. Part of his motivation had been to understand why nature had made the "extra electron," the muon. We still do not know the answer to that question. But on the other hand, his discovery has had the effect of changing the subject. Nowadays, rather than asking Why is there an extra electron? we ask Why are there only three such particles? And in answering *this* question, a striking clue is available. It pertains to the quarks, the particles of which neutrons, protons, and the like are made. Present knowledge holds that there are six such quarks—but also that they come in pairs. So there are three pairs of quarks, just as there is the electron, the muon, and the tau.

But Perl is not too interested in the tau any more. Right now his burning passion has to do with the quarks. He wants to find one.

The remarkable thing about quarks is that, while the indirect evidence for their existence is persuasive, the direct evidence is positively nonexistent. In spite of decades of effort, quarks have steadfastly eluded all attempts to observe them. So high-energy physicists find themselves in the uncomfortable position of accepting the existence of particles they have never seen. Perl would like to change that.

This new experiment is something of a departure for him. It is not big science but small science—ultra-small science. The whole experiment occupies no more than a tabletop, and it is being conducted by Perl with help from precisely one assistant. The method is to squirt tiny droplets of water out of an injector, suspend them in midair, and search for any quarks that might have become imbedded in them.

When I visited Perl, he was wrestling with those water droplets. His injector was not making them correctly. The droplets were not of uniform size. After decades of writing grant proposals, managing large collaborations, and participating in the construction of immense instruments, Perl was now back to what many people would regard as the very essence of science: he was incessantly turning over in his mind a single technical problem. He wandered into people's offices to ask them questions, drew pictures on his blackboard of how his injector worked, gazed at it, and thought.

One day I arrived at Perl's office to find him in conference with his assistant. They had gotten hold of an ink-jet printer, standard companion to many home computers, and had taken the thing apart. The working heart of such a printer is not unlike their injector: a device that squirts a tiny jet of ink at the paper one is printing. Perl was wondering if the engineers who had built it had not perhaps already solved his problem for him.

I stood silently at the door and listened. Perl and his assistant were gazing at the device through a magnifying glass. They were trying to figure out how it worked. They were discussing the function of each component part. They talked about cutting the thing up with a hacksaw to take a look inside.

Standing there in the doorway, I was struck at how prosaic was the scene before me. Nothing of what I saw looked like the popular image of research at the very frontiers of knowledge. Nothing smacked of Great Minds Wrestling With Great Ideas. Two people sat beside each other at the table, speaking quietly, turning over and over in their hands the tiny device. Down the corridor, idle right now, sat a potentially epoch-making experiment.

The quarks for which Perl is searching, the tau he discovered, and all the other elementary particles fit into a theory known as the standard model. This theory has proved remarkably successful. While it does not address every feature of elementary particle physics, those features it does address it does so correctly. Most physicists are happy to have such a successful theory in hand. But not Perl. He is deeply unhappy with it.

In explaining why, Perl is fond of quoting Francis Bacon:

. . . they are ill discoverers who think there is no land when they can see nothing but sea.

Concerning this epigram, Perl has written,

The quotation describes the standard model of particle physics, a uniform and endless sea which seems to surround us. Perhaps the tau will provide the island, the new land, which will enable us to climb out of that sea.

The image that Perl has picked from Bacon's epigram is one of oppression. The standard model "surrounds" him in a "uniform and endless" featurelessness. He is looking for some way to pass beyond this—to pass into a new land, into a whole new continent of discovery. What to most physicists is a pleasing agreement of theory with experiment is to Perl an oppressive blankness.

Martin Perl is dissatisfied with the standard model, and he is dissatisfied with the present state of physics, and he is dissatisfied with himself. When I visited him, he was sixty-six years old. That is an age at which most people are winding down their careers—but Perl is worried that he is not doing enough. That is an age at which most people are consolidating past gains—but Perl is restlessly searching for more.

He sees high-energy physics nowadays as being at a standstill. "The whole field is coasting," he says, "and it has been for years. Not much truly exciting is going on. My guess is that this is why so many people were so desperate to get the Super Collider. They were hoping the machine would show them the way to new science. The person who finds a flaw in the standard model, the person who finds the direction to the next step in the stairs that leads us beyond the standard model—that person is going to be famous."

Perl leaned forward in his chair. "I want to make a difference," he said. "It is very important for me to have an impact on physics. Very important to contribute more than just some little dotting of an *i* or crossing of a *t*." This was the discoverer of the tau particle speaking to me, not some young and ambitions hotshot who has yet to prove himself. This was a man at the peak of his profession, a man who, though he did not know it at the time, was within months of receiving the Nobel Prize.

Martin Perl leaned back and smiled. There was wonder and admiration in his voice. "How do they do it, these *really* smart people?" he mused. "How do they keep on having so many good new ideas?"

8

Our Address in the Universe

Margaret Geller and John Huchra

One day in the summer of 1985, there occurred an event of some importance in the history of science. Not often can the precise moment of a great discovery be pinpointed: normally, they emerge piecemeal, and slowly. But a case can be made that our knowledge of the universe underwent a profound transition at a definite moment. It was the instant a graduate student at Harvard picked from the output of a laser printer the figure reproduced on page 118. Her name was Valerie de Lapparent. With her advisers Margaret Geller and John Huchra, she was engaged in mapping our corner of the cosmos. It was part of the work for her Ph.D. thesis: the figure was their final result. She walked into their offices and laid it on their desks.

Galaxies, enormous swarms of stars, are the fundamental units of construction of the cosmos. They are what the universe is made of: they are to it as islands are to an archipelago. Geller, Huchra, and de Lapparent were studying how the galaxies are distributed. They were studying the shape of the archipelago, if you will. Prior to this moment, astronomers had known—or rather, "known"—that galaxies scattered more or less uniformly through space. Everyone knew the universe was uniform. Geller and Huchra knew it.

But in their new map the distribution of galaxies was anything but uniform. On de Lapparent's figure lying before them on their desks, the galaxies lay congregated in extraordinary structures. It took some time to sort out the nature of these structures. We now know they are sheets. The sheets are not flat. They curve. The galaxies are aggregated into mighty arcs, the arcs curling about and intersecting one another in a complex, irregular pattern. Surrounded by the curling arcs are voids, regions that contain no galaxies whatever on their map. The new image of the universe provided by their research bears some resemblance to a foam of soapsuds, or to a sponge. Geller and Huchra looked down onto their desks and beheld a sudsy universe.

Extending entirely across their map lay the largest structure anyone has ever seen. It was an irregular sheet of galaxies that has since been dubbed the Great Wall. Down toward its base, the sheets coalesce into a structure by happenstance reminiscent of a person—a stick figure with bow legs and sinuous arms. It looks somewhat like those ancient, enigmatic pictographs one finds etched on sandstone walls in the American Southwest.

It is difficult to appreciate the scale of Geller and Huchra's map. On it cities, states, even continents are too small to be noticed. The very Earth itself is shrunk into invisibility—it is smaller than an atom on their scale. Astronomers customarily reckon distances to things in terms of the time required to reach them when traveling at a certain speed. It is like saying Los Angeles lies five hours from New York by air. In the astronomical context the relevant speed is the greatest of them all, that of light. A ray of light, capable of circling the Earth in a fraction of a second, takes eight minutes to reach the Sun, and four years to reach the nearest star. Many of the stars visible to the naked eye are so distant that the light we see them by was emitted during the Middle Ages. Our galaxy, the swarm of stars in which we lie, is so enormous that when astronomers study its distant edge they are observing 100,000-year-old light.

But on Geller and Huchra's scale, even these distances recede into insignificance. In their map of the universe, the galaxies themselves shrink into mere points. The ancient light they surveyed was emitted far back in the geologic past. The stupendous yawning gulfs that are the voids are hundreds of millions of light years in extent. The Great Wall is at least half a billion light years long, and may well be longer—we have never found its edge.

But as John Huchra gazed down at their new map, his thoughts were not of the inexhaustible grandeur of the cosmos. "I looked at that thing," he told me, "and I panicked. My only reaction was, 'Oh, my Lord, I must have made some terrible mistake.' And I spent the next month checking my data, to find the error."

There was no error. As for Margaret Geller, the thought had never crossed her mind: she sat right down with de Lapparent to begin interpreting the results. "She had more faith in me," says Huchra with a grin, "than I did myself."

Huchra grew up in Ridgefield Park, New Jersey, a blue-collar community just over the George Washington Bridge from New York City. His father worked as a freight conductor for the rail-road. Poverty was an ever-present threat: at one point his father was injured on the job and out of work for months, and with neither pay nor workman's compensation coming in, the family survived only through the help of handouts from relatives and friends.

"I played a lot of softball in those days," Huchra told me, "and spent a lot of time in a nearby swamp. Some friends would trap muskrats there, and raise money by selling their pelts; I collected bottles and turned them in for the deposits. I recall that on my seventeenth birthday I was walking down the street and passed a snow fort built by two other kids. They threw a few snowballs, one thing led to another, and eventually the mother of one of them called the cops. They came and discreetly put an end to things.

"I was exceedingly skinny as a kid—seventy pounds as a freshman in high school—and subject to bullying. Finally, my father got the point and gave me a few lessons from his Marine Corps training. Soon after, walking home from a Confraternity of Christian Doctrine class at church, I saw two kids waiting for me down the street. They were members of a gang who, in groups of five or so, would beat me up each and every Sunday after church. This time, instead of trying to avoid them, I walked up and out-and-out attacked: straight-armed one into a telephone pole and knocked the other over a hedge.

"One day in gym I was paired against a hood at wrestling. All the guys in class were rooting for him. I had my opponent pinned to the mat in two minutes, thirty seconds. Much astonishment all around. Soon there was a rematch to correct what had gone wrong, and this time I had him pinned in thirty seconds."

Huchra joined the wrestling team and the track team, and he did well. It would be pleasing to report that as a result of all this, he gained a certain amount of respect from the bullies. More to the point, however, was that he became friends with a number of athletes: "When you are friends with a 240-pound shot-putter, the hoods leave you alone."

Although both his parents were intelligent, neither had more than a high school education. They could not help him with his homework. Nevertheless, Huchra did well in school. He loved science and, like most who go into the field, knew more about it than his teachers. Most of his classmates went on to vocational training: he went to MIT. "My parents broke their backs to help financially," he told me, "and I managed to snag some high-paying jobs in construction, driving trucks and the like. One summer I made $2,500 loading trucks at a freight dock while my classmates, all children of doctors and lawyers, were pulling in maybe $400 as summer-camp counselors."

As the child of working-class parents, Huchra felt somewhat out of place at college. He did well in terms of grades, but nevertheless sometimes felt inferior to the other students. He was often nagged by the feeling that they possessed some indefinable spark that he lacked. As a senior, he never mentioned to his adviser that he was applying to graduate school: he did not think he was good enough to get in. Although Huchra liked science as a student, and wanted to become a scientist, he would not have been utterly devastated had he been rejected. By the time graduation rolled around, he had been offered a job as foreman and shop steward at a trucking company, a job he would have been pleased to accept—the pay would have been good, and the position one he respected. Even when he did get into graduate school, and continued doing well, the feeling persisted.

Margaret Geller grew up in Morristown, New Jersey—not that far geographically from Huchra, but socially light-years distant. Her father was a solid state physicist at Bell labs. She remembers with pleasure trips to his laboratory, where he would show her how to measure X-ray diffraction patterns to determine the spacing of atoms in crystals. Margaret would enter the numbers into a wonderful old mechanical calculator, which would bang and thump mightily as it ground through the calculations. People from the lab were continually dropping by to say hello.

Although she played with other children, Geller was continually beset by a sense of loneliness. In looking back on these times, she sees herself as somehow different, and as a child who had few close friends. School was boring, and an oppression, and she would often feign sickness in order to escape. Her parents soon recognized what was going on, and they let her remain at home provided that she study. Geller remembers with distaste a particular teacher who would have done serious damage had not her parents been so supportive, a teacher who publicly ridiculed her for being shy and awkward. She *was* shy, and she felt she was awkward.

And she was smart. A smart girl, good at science.

While I was preparing to write this book's first chapter on Cecilia Payne and Annie Cannon, I did a good deal of reading on the status of women in science in their day. What I found horrified me. Even though I knew that women scientists had faced terrible discrimination, I was still unprepared for its magnitude and blatant hypocrisy. But throughout it all, I was able to console myself with the thought that this disgraceful discrimination was a thing of the past.

Things have gotten better since those dark ages. Women astronomers are allowed to use telescopes nowadays. The courses they teach are listed in college catalogues. And an administrator would be fired were he to declare, as Harvard's president had, that no women would ever have a position at his university so long as he lived.

And yet—when I began preparing to write this chapter, I learned to my dismay that in some important ways very little has changed. Females still face discrimination in math, science, and engineering. They face it at every stage of their lives: when they are kids in elementary school and young women in college, when they are beginning their careers and at the most senior level. Nor is this discrimination reported by a small number of malcontents: it appears to be a persistent and widespread pattern encountered by great numbers of women. There is a structural component to the problem as well. It is systematically harder for women to practice science than men. Our society has erected a set of barriers blocking their paths. These barriers are complex and pervasive, and they require women scientists to make heroic sacrifices—sacrifices that men are more rarely required to make.

The net result is a shockingly small number of women practicing science today. A recent study has revealed that in the United States a mere 3 percent of physicists are women.

When I learned of this statistic, I was astonished. Perhaps naively, I had thought that the number of women in science was far higher. Everywhere I looked, I saw programs designed to increase their number: federal programs, efforts in colleges and universities, special incentives and summer camps. Also, everywhere I looked, I saw people declaring the participation of women in science to be a Good Thing. Nowhere did I find anybody declaring it to be undesirable—the old attitude that a woman's place was in the home had gone the way of the horse and buggy. And finally, all this social pressure for change seemed to be working. In the media I regularly encountered reports on the achievements of female scientists. Newspaper articles and television documentaries regularly highlighted their work. What I had not realized was that this media attention is selective: with the best will in the world, the reporters are seeking out the relatively few women scientists for their accounts.

My biggest surprise came when I learned of the results of a study of other nations. A colleague who knew of the study asked me to guess the ranking of four countries in terms of their representation of women in science. The countries she chose were Turkey, Italy, Canada, and the United States. My guess was that Canada and the United States, with their strong women's movements, would have a relatively large number of women scientists. Turkey and Italy, on the other hand, with weaker women's movements, would have smaller numbers. But as it turned out, I had guessed exactly wrong. The percentage of practicing physicists who are women in Turkey and Italy is nearly *ten times* higher than in the United States and Canada. Leading the list in terms of representation was a nation I never would have thought of—Hungary, nearly half of whose physicists are women.

People speak of a pipeline, carrying students from their early schooling on through college and into their professions. In the case of women going into science, the pipeline is leaking at every seam. In 1977 there were 730,000 girls of high school sophomore age enrolled in science or engineering courses throughout the United States. By 1980 they had become college freshmen—but fewer than half of them had elected to continue their science studies. The rest had chosen never to take another course in the field. Four years later, those young women had graduated: out of the initial crop, only 28 percent had obtained degrees in science or engineering. And by 1992, at which time they were thirty years old, a mere 1.3 percent of them had obtained graduate degrees in the field.

But why should everyone taking high school science go on to professional training? It would not even be a good thing—were they to do so, the field would be overpopulated in an instant. That pipeline had better be leaky. The real point is that the leaks are far bigger for women than for men. Eighteen times more mathematically talented boys go on to earn Ph.D.'s in the physical sciences than girls of fully equal talent.

Something is driving women out of the field. They are leaving in droves, and they are leaving at every stage of their lives. The most distressing story I know of in this regard concerns a program Wellesley College started up, designed to attract its women students into science. The program was ambitious, and it succeeded beautifully. As time passed, more and more of their students graduated with degrees in science. But, in a follow-up study, people at Wellesley learned to their dismay that the effects of their program were in many cases short-lived. Fully 20 percent of their science students had changed their minds, dropping out of the field within a mere *six months* of graduating.

Why? What is happening? In spite of all the media attention, all the special programs and the rhetoric and the best intentions, why is the representation of women in science so dismal in this country? There is no single answer to this question. Rather, there is a whole network of answers.

When my wife was a little girl, she loved arithmetic. She was good at it, and she would do it for fun. But when she was about eleven, all this changed. She developed a fear of mathematics. Her mind would go blank when anybody tried to explain it to her. She does not recall any particular reason for this—no cruel teachers telling her she was acting wrongly, no boys sneering at the girls. These were what Cannon and Payne faced, but they do not happen so very much any more. Rather, my wife's transformation simply seemed to happen by itself. But it had not happened by itself. It happened for a reason.

Not long ago, the makers of Barbie, the doll beloved of little girls, marketed a model that said "I hate math" when squeezed. There was a furor, but Barbie was merely reflecting a social reality. No matter how we may wish it were otherwise, there is a widespread myth in our society that girls are "intuitive" and "artistic," while it is the boys who are "good at math." The teachers of math and science that children encounter are more likely to be men than women. Conversely, their female teachers are likely to be poor at these skills. The nurse at the doctor's office is apt to laugh deprecatingly that she is "no good at numbers," while the auto mechanic is just about sure to be a man. The parents of a young girl may give her a model train set for

Christmas—but more likely than not it is daddy who sets it up, while mom busies herself in the kitchen.

In one study, impartial observers sat in on math and science classes, and they carefully watched the behavior of the teachers. What they found was deeply unsettling. The teachers called on the boys more than on the girls. They praised correct answers from the boys, but did not do so when a girl got a problem right. They would fail to notice the upraised hand of a girl, waiting instead until a boy raised his. They would attribute the correct answer to a question to a boy, although in fact it had been given by a girl.

This behavior was seen both in male and female teachers. Most unsettling of all, the teachers were not even aware of their behavior. When directly queried, they would say that they valued the girls' contributions, and they were horrified to learn of what the observers had found.

Children hear our rhetoric about the equality of men and women—but they are also extraordinarily good at hearing what we do not say out loud, but mean all the same, and at picking up on the subtle cues that surround them. These cues are insidious, and they are all the more potent for being subliminal.

They extend into adult life as well. Not long ago, an experiment was performed in which professional mathematicians were given a research paper in mathematics and asked to rate its quality. Half the responding mathematicians were given one paper, half another. Unknown to the respondents, however, was that the two papers were identical. They differed only in the name of the purported author: half were by "John T. McKay" and the other half by "Joan T. McKay." The results of the study were that the papers "written" by the man received significantly higher ratings than those "written" by the woman—even from the female mathematicians.

Not long ago, I attended a committee meeting at which two students were present. The person sitting next to me listened attentively as the male student spoke—but when the female began speaking, nudged me to point out the pretty view through the window. Furthermore, this person, who valued the woman student's comments so lightly, was herself a woman.

One of the most respected members of my own astronomy department is a woman. She is often interrupted when she speaks at our meetings. If she wants to make a point, she is sometimes forced to contend with a barrage of disruptions, disruptions that the male faculty do not face. And here, too, I am convinced that this propensity to interrupt is unconscious. If anyone were to ask the interrupting speakers, they would praise this woman as one of the stars of the department.

In these and countless other ways the message is sent—subtly, subliminally, but all-pervasively—that what women say is not worth listening to. Furthermore, the woman who contends against this attitude is placed in a difficult situation, for her very effort to do so is often turned against her. The man who does not allow himself to be interrupted we praise as being powerful and confident, but if a woman does this, we are likely to condemn her as being "pushy" and "overly aggressive." Indeed, many of the qualities that we praise in men we condemn in women. Competitiveness, aggressiveness, the drive to get ahead—these are popularly regarded as masculine traits, but inappropriate for a woman.

But this overlooks the fact that plenty of women are competitive—and plenty of men are not. Furthermore, we seldom stop to ask whether one *needs* to be competitive to succeed at science. In my opinion, one of the greatest reasons to hope that more women will go into science is that their presence might help break down this pernicious attitude. Yes, you have to be aggressive and competitive to win at football—but why are these so important if your dream is to unlock the secrets of nature?

Is it all true? Could it be that there is a genetic difference between men and women that makes men biologically better at math and science than women? Unpleasant as the thought may be, is it possible that all these popular myths are actually correct?

In fact, studies have shown that young girls are just as good at math as boys. Girls' average math scores on the Scholastic

Aptitude Test are no different than boys'. By age seventeen, however, the boys' scores begin to pull ahead, and thereafter those of the girls never catch up. But is this the result of a difference in genetic makeup, or is it a product of social conditioning? There has been study after study on this question, and at present there is no general consensus as to the answer. We simply do not know.

But I would argue that we are asking the wrong question here. For in many ways the answer simply does not matter. Even if, *on the average,* men were better at math and science than women, there would still be plenty of highly able women—more than enough of them to fill our math and science departments. Furthermore, the real problem here has nothing to do with intrinsic ability. It has to do with what people think about this ability. In one study, a group was assembled of boys and girls selected to be equally good at math. They had *the same* math SAT scores—but the boys regarded themselves as being better all the same. Most distressing of all was that the girls agreed with them: they, too, regarded the boys as being better at math than they. In another study, the career choices were tracked of boys and girls selected to be equally good at math. Roughly one-twentieth as many girls as boys from this group went on to become professional scientists. And as for the rest of the girls, good mathematicians all, they went on to do . . . what?

The greatest disparity between the representation of men and women in science is at the very highest levels. A fairly large percentage of junior scientists are female, while only a small fraction of the full professors are. One might suppose this is because the influx of relatively large numbers of women into the field is a recent phenomenon, and they have not yet had time to reach the senior positions. Unfortunately, this is not the case. The disparity has persisted for so long that they would have done so by now, had they not been actively prevented from doing so.

Rather, people speak of a "glass ceiling," a subtle yet all-pervasive discrimination that prevents many women from climbing all the way to the top of the ladder in their professions. Older women speak often of having encountered it. When an important discovery is made, the news that races around the world via e-mail and telephone calls somehow fails to reach them. They find themselves mysteriously not placed on the most important committees, not asked for their opinions on the most important decisions that need to be made. And as reported by the journal *Science:*

> There is a long history of women being invited to talk at math conferences only after organized protests. In 1978 no women (and approximately 100 men) were invited to talk at the prestigious International Congress of Mathematicians, held every 4 years. Before the next meeting, in 1982, women protested, and a few were invited. In 1986 only one woman was invited and she spoke on the history of math. "We didn't make a fuss, and here they go again," recalls Linda Keen, president of the AWM at the time. Keen organized a protest, and, according to Marina Ratner, Berkeley's only woman mathematics professor, "at the last minute three women were grudgingly invited." Ten women were invited in 1990 after the American delegation "reminded" the organizing committee to invite women.

In a celebrated case, a tenured full professor of neurosurgery resigned after twenty-six years at Stanford University, citing "pervasive sexism" as her reason.

So—is there a genetic difference between men and women? In her inimitable fashion, Margaret Geller puts it this way: "Of course there is! Good heavens, it would be a pretty boring world without that. But this has nothing to do with the problem."

The Second Shift is a book I found particularly disquieting. Its author, Arlie Hochschild, is a sociologist concerned with the structural problems faced by women who pursue careers in science. The second shift is what you do after you have finished with your "real" job. It is a second career, just as demanding as the first, that commences at the end of the day when you are most tired—and you come home to the kids.

Hochschild studied a group of two-career families for a number of years. Her book is a report of what she found—and what she found is exhaustion, people working at the very limit of their strength to maintain good homes for their children while pursuing demanding, full-time jobs at the office. Furthermore, she found that this burden falls preferentially on women. Even in families in which both partners have agreed to share the household duties equally, she found that more often than not it is the women who are doing the bulk of the child care. She writes of women making enormous sacrifices to maintain their careers—sacrifices that of course many other professional women have made, sacrifices that while difficult are certainly not impossible, but sacrifices that their male colleagues are called upon to make far less often than they. She writes of her women students contemplating their futures with trepidation. These are young women who intend to go into some demanding career, but who grew up witnessing the heroic efforts their mothers made to balance their first and second shifts—and who are deeply concerned that they themselves not be forced into such exhausting lives.

Hochschild's study is anecdotal rather than statistically rigorous. We have no data demonstrating that in two-career families nationwide, it is the women who make greater sacrifices than the men. But we do have other data bearing on this issue. One statistic is that irrespective of who does what task, women are more likely than men to be partners in two-career families, with all these families' attendant burdens. Statistically, when a woman scientist marries, she is more likely to marry another professional than is a man when he marries. A good number of male scientists have wives who have opted out of a career, choosing instead to stay at home to raise the children. But very few women have husbands similarly free to relieve them of the burden of the second shift.

The consequence is that marriage for women professionals is a far more demanding business than it is for men. And the consequence of *this* is that many women decide that their only option is to choose—marriage or career, but not both. One result is the relatively low number of women in science. Another is the relatively low number of women scientists who marry. Nationwide, only 18 percent of all male chemists are single—but a whopping 38 percent of all female chemists are.

Some time ago, a female physicist from England moved to Italy to take up a job. "I have light brown hair, but there I was considered a blond," she says of the experience. "I don't consider myself particularly attractive, but I couldn't walk three feet without men pestering me and asking me for my phone number." Nevertheless, the percentage of physicists who are women in Italy is nearly ten times that in the United States. How can this be?

The reasons are largely structural. One is that Italian girls are simply not given the chance to drop out of math and science. Courses in these subjects are required of everybody through high school. Furthermore, the teachers of these courses are likely to be better prepared than in the United States. One of the most embarrassing things about early math/science education in this country is the number of teachers who are simply no good at these subjects. They were trained in departments, not of mathematics or chemistry or astronomy, but of education. But in Italy the high school math and science teachers learn their subjects in a more rigorous way. They are required to take the same classes as those students going on to careers in research. This also means that since a good number of these teachers are women, the population of women in these courses is fairly large. In the United States, in contrast, women in math or science classes are often rare, and they stand out to their embarrassment.

Most scientists are employed in colleges or universities, where their jobs are protected by the institution of tenure. Tenure is a commitment on the part of the employer to guarantee employment, barring the most extreme of circumstances, for the remainder of the holder's active life. This is an extraordinarily expensive commitment on the part of the employer, and the most stringent requirements are set forth for attaining it. The applicant for tenure must demonstrate true excellence in research as well as teaching. And the only way to demonstrate this is to work—hard.

The applicant for tenure must conduct research, and lots of it. The problem is that this pressure to "publish or perish" comes when the applicants are in their early thirties in the United States—and this is just when they are having children, and rearing them through their most demanding early years. Just when women scientists are most vulnerable professionally, just when they must work hardest to achieve tenure, is when the pressures of rearing a family are also greatest.

It's a bad juxtaposition. One option is to delay the tenure decision: to remain in a nontenured junior status for a number of years by going part-time while having babies and seeing them into kindergarten. The problem with this option is that it carries a stigma in America. "She can't be very serious about her career if she's going part-time like this," we are apt to say under our breaths as we labor in our laboratories, gazing enviously out the window at our colleague and her kids, off on a jaunt to the neighborhood playground. But in Italy there is no such stigma. Notorious baby-lovers that they are, Italians are perfectly respectful of such an option, and many women there avail themselves of it. Nor is this happy attitude confined to that one country. One junior woman scientist reports having spent time in Israel, where she was actually chided by the head of her research group—a highly competitive and active group, by the way—for *not* taking the day off on her little boy's birthday.

The dismal representation of women in mathematics, engineering, and the sciences in our country is not going to change until we address these difficulties they face. It will not be an easy task. On the other hand, the numerous studies that have been done at least tell us some of the things that need to happen.

The Italian government provides free day care to every working woman. Our government should do the same. Day care centers should be an integral part of every workplace: every university, every observatory, every conference. Funding agencies should routinely allow day care costs as a legitimate research expense. They already pay for equipment and travel to scientific meetings: they should also pay for nannies.

Pre-service teachers should be given the same training in math and science as those students going on to other fields— the women as well as the men. Girls should not be allowed to opt out of these courses in school. Colleges and universities should encourage junior faculty, the men as well as the women, to work part-time during their children's early years.

Structural things such as these are relatively easy to tackle— at least they are visible. But how can we alter the invisible, subliminal messages our culture sends to women and girls? How to do something about prejudices of which people are not even aware? It may seem impossible, and yet all in all I am not entirely disheartened. For I am sure that most of the teachers who learned to their dismay that they had been treating the boys better than the girls were able to alter their teaching practices. In the final analysis our rhetoric, our best intentions *do* matter. What we say we wish is usually what we want to wish, and if we are made aware of our subliminal discrimination, perhaps we can change. All it takes is the will.

Margaret Geller went to college at Berkeley. Although there were not that many women in the science courses she took there, she felt encouraged. Not so in graduate school, however. It was an ivy league university that had only recently gone co-educational, and there were people, both faculty and other students, who said things about the role of women in science that were deeply shocking to her. Although she did well academically, she emerged from graduate training with her self-confidence badly shaken.

She took a job at the Center for Astrophysics, a joint venture between the Smithsonian Institution and Harvard. One year later, Huchra arrived. Geller's degree had been in physics, Huchra's in astronomy, and much of her expertise in astronomy she got from him. She feels she was a slow starter, partially because of her bad experience in graduate school. "It was not until, after three years at Harvard, I took a visiting post at Cambridge University in England that I really sat down and thought about what I wanted to do. I felt I had been just diddling around up to then. I didn't feel I had developed my own style. I did not have a program, an overarching research thrust. When I was in England I decided that now was the time to put up or shut up—that if I wasn't going to make a really serious commitment to the field, I might as well get out. And if I was going to be in science, the exciting thing was to be *out there*.

"I began to think about what was known concerning the large-scale structure of the universe. I realized that, in fact, very little was known for sure. People were generalizing from a severely limited number of observations. I recognized just how fragile were the foundations of the field. And when I returned to Harvard, John and I sat down and mapped out a long-term research effort aimed at getting at these issues."

The problem of mapping the universe is radically different from that of mapping the Earth. In some ways it is easier. Columbus could not see the new world from Spain—he had to come here to discover it. But there is nothing in intergalactic space blocking our view of the distant cosmos. The universe is of a transparency exceeding that of the most exquisite glass. To map it, all we need do is look.

The hard part does not have to do with clarity. It has to do with perspective. The view we have of the universe is two-dimensional. The nighttime sky appears, both to the naked eye and to the astronomical telescope, as if it were pasted on the inside of a dome. How far away are things? It is hard to find out.

In daily life, clues indicating distance are everywhere. We judge an automobile with brilliant headlights to lie closer than one whose lights appear dim. Similarly, the more widely separated the headlights, the closer the vehicle. But both these judgments rely on assumptions—that headlights are all of the same brightness, and that they are mounted pretty much the same distance apart. Assumptions such as these are valid in daily experience, but in the context of the astronomical universe they might not be. A brilliant star need not be closer than a dim one.

It might be at the same distance, but more luminous—it might even be farther away. Similarly, a large galaxy, mistakenly thought to lie close at hand, might in fact be a distant giant.

The errors introduced by the misjudgment of distance in astronomy are legion. Most constellations, for instance, have no objective reality. They are illusions. The various stars of the Big Dipper only appear to lie close to one another. In reality, they are at radically differing distances from us, and therefore from one another. Among all the obvious groupings in the sky, only two, the Pleiades and part of the constellation of Orion, are genuine structures.

As I was beginning my career as an astronomer, a poster was produced representing what the sky would look like if only we could see the distant galaxies with the naked eye. I kept a copy on my office wall, and I always regarded it with a mixture of exhilaration and frustration. There were tantalizing hints of a pattern visible. The distribution of galaxies was irregular, blotchy. Faint knots stood out, regions where they clustered together. Other regions were relatively devoid of galaxies. Poorly defined strings appeared to wander this way and that through the pattern.

But the predominating impression conveyed by the poster was of a rough uniformity in the distribution of galaxies. Aside from the clusters and other faintly defined structures, they appeared to scatter more or less randomly through space. It was just such data, in fact, that had always led astronomers to believe that the universe was essentially uniform. On the other hand, I was at the same time vividly aware of the limitations of this conclusion. If the galaxies, themselves so enormous, aggregated into yet larger structures, these could easily be missed in such a view, which looked *through* them, the structures superimposed, melded together and lost in a general confusion.

Did the patterns on the poster possess the same illusory character as do most constellations? Or were they real? I was never able to decide—and neither was anybody else. The most detailed of tests, the most sophisticated of mathematical studies, proved unable to extract reliable information. Lacking the all-important dimension of depth, there was little anybody could do. Standing before the poster on my wall, I experienced the exhilaration and frustration of living in a universe without perspective.

Geller and Huchra mapped the cosmos by finding the distances to the galaxies. This they accomplished by inverting Edwin Hubble's famous discovery of the expansion of the universe. Hubble had announced in 1929 that each galaxy is receding from us, with a velocity proportional to its distance. But if velocity is proportional to distance, then distance is proportional to velocity. You can find the distance by finding the velocity.

The velocity, in turn, can be found from a galaxy's redshift. The redshift of a galaxy's light—the shifting to the red of its constituent colors—arises for the same reason that the pitch of an automobile's horn drops as it passes. Both are instances of the Doppler effect, and both can be used to measure the velocity of motion: the greater the shift, the greater the velocity.

Redshift is measured by passing the light of the galaxy through a device such as a prism, which spreads it into its constituent colors, recording them on a photograph, and then comparing the colors to a reference standard. The process is easy to describe, but in Hubble's time it was fiendishly hard to carry out. In the 1920s and 1930s long time exposures were required, even using large telescopes, to gather enough light from a distant galaxy to expose the photograph. Hunched over the eyepiece for hours on end, guiding the telescope to keep the image of the galaxy trained on the measuring instrument, Hubble and his colleagues might spend an entire night measuring the redshift of a single galaxy. Exposures lasting several nights were not uncommon. The world's record was a ghastly eight nights of work spent on a single particularly faint galaxy: when the photographic plate obtained on that run was developed, it turned out to be useless. Such failures were common.

Hubble worked at the 100-inch telescope on Mount Wilson overlooking Los Angeles. It was reached by a toll road in those days, a road made of dirt. It was so narrow that two cars could not possibly have passed one another—traffic ran up the mountain in the mornings, and down in the afternoons. Sometimes an astronomer would hike to the summit via a trail. So enormous were the telescopes needed to observe the distant galaxies, so exacting were the technical requirements, that throughout the entire world there were no more than half a dozen people engaged in carrying on the work.

At the time my poster showing 1 million galaxies was produced, redshifts were known for less than one-tenth of 1 percent of them. It was the sort of situation that makes a good observer's blood boil.

The past decade has witnessed a quiet revolution in the technology of capturing light for astronomical purposes. In the old days, things were done photographically. Nowadays, people do them electronically. Light from a telescope is focused onto a solid-state device that generates a minute electrical current when illuminated: the current is amplified and displayed as an image.

Such devices vastly simplify the problem of obtaining redshifts, and they are incomparably more sensitive than photographic film. The marathon eight-night effort that in the 1930s had failed to measure the redshift of a faint galaxy can now be duplicated—and successfully—in a matter of minutes. When Huchra arrived at the Center for Astrophysics, he joined a group led by Marc Davis engaged in building an instrument to accomplish this task.

The device Davis's group ended up building was a cylinder, 10 inches in diameter by 2 feet long. It was light enough to be carried by one person. Not coincidentally, it just fit beneath the seat of a commercial aircraft. They called it the Z-machine, Z being the astronomical symbol for redshift. Once built and running, it was Huchra who used it to take most of the data.

John likes to say that the happiest times of his life have been spent on mountaintops. For the redshift survey he used a telescope situated on Mount Hopkins, outside of Tucson. As telescopes go, it is relatively small, but when hooked up to the Z-machine, it is incomparably more powerful than the giants with which Hubble had worked.

When observing, Huchra sleeps during the day at a residence for astronomers on the mountain. Rising in midafternoon, he carries a bag of groceries to the telescope dome and cooks his dinner there, sitting outside to watch the sunset as he eats. About this time, he will open the slit of the dome, to allow the telescope to cool to match the outside temperature. He wants to minimize the flow of warm interior air out the slit during the coming night, which would cause the sky to twinkle—a pretty sight, but a perennial problem for astronomers, who need the crispest possible view of the heavens. At twilight, he prowls the dome, checking the electronics—often there are glitches, and there is much scurrying around to be done. Once it is fully dark, he uses the Z-machine to observe a star whose redshift he knows, testing to make sure things are working properly. Then begins the first real observation.

Huchra has arrived armed with a stack of index cards. On one side of each card is a photograph of a region of the sky. Bright stars and dim are scattered across the photograph—and galaxies too, each indicated by a tick. On the reverse side of each card, he has written the galaxies' celestial coordinates. His task is to get the telescope pointed directly at each target, so that its light may enter the Z-machine's entrance slit. It's a matter of marksmanship.

In contrast to stars, which are individual points of light, galaxies appear as diffuse regions of faint, pearly luminescence. Only slightly brighter than the surrounding empty sky, they can be hard to make out even through a telescope. Over the years, Huchra has evolved a variety of tricks to move things along. During the evening, he will have been careful to avoid all bright lights, in order to dark-adapt his eyes—a single glance at a fluorescent tube would seriously degrade his eyesight. While observing, he might jiggle the telescope slightly, finding it easier to spot the galaxy if it suddenly moves.

Finally, when the target is acquired, there will be a wait of twenty minutes or so while the Z-machine gathers the data. Geller has been out to the telescope on occasion, but she does not really like observing. To Huchra, to two full-time employees stationed in Arizona, and to occasional graduate students working on their dissertations, falls the burden of making the obser-

vations. The work is exacting, but John finds it peaceful and ful-
filling. Inside the dome, all the lights are off. Starlight filters
through the open slit overhead. The telescope is a massive affair
of struts and girders silhouetted against the sky: a quiet hum
can be heard from a clock drive as it steadily swings the instrument to follow the rotation of the heavens. While he waits,
John might take a stroll outside to stretch and check for clouds,
and to catch the nighttime smells. Inside, he might watch television, or play some tapes—Brahms, rock, whatever.

The first use of the Z-machine was to conduct a broad survey of
galaxy redshifts. Once this was completed, Marc Davis left to
pursue other problems in cosmology. As a result, Davis, who
played a pivotal role in getting the redshift survey going, missed
out on its most spectacular discovery.

There is always an element of luck in science. Most discoveries cannot be credited to a single person. Sometimes I think of
research in terms of an analogy. I think of it as the construction
of newer and ever more sophisticated vehicles, and the use of
these vehicles to explore unknown territory. It is always a question of strategy to guess what type will be best suited to whatever terrain happens to lie over the next hill. Should we put
together a land rover? An amphibious vehicle? Davis's initial
survey was designed to study certain mathematical questions
relating to the distribution of galaxies, and it succeeded at that
task. Unfortunately, however, his design turned out to be ill-
suited to the discovery of the structures that Geller and Huchra
later found—structures nobody could have known in advance
were present.

In 1983 Geller and Huchra sat down and designed a survey
specifically tailored to the study of large-scale patterns in the
universe. To illustrate the strategy they settled on, Geller uses a
particularly vivid analogy. She takes a map of the Earth, and she
cuts from it a single, confetti-like strip.

She likes to emphasize that the volume of the nearby universe—the portion that we have even begun to map systematically—bears the same relation to the entire observable universe

as does the state of Rhode Island to the Earth as a whole. Suppose, she argues, that some hypothetical extraterrestrial being wanted to learn something about the geography of our planet, but was only able to survey a single, randomly chosen region the size of Rhode Island. How well would the being do? How accurate would be the map of our world that it drew?

Most of the world is ocean, so very likely the extraterrestrial mapmaker would end up concluding the Earth consists of nothing but water. If, on the other hand, the region surveyed happened to fall on land, it would probably miss the coastline, and so give no hint of the existence of oceans. The patch surveyed might fall in a desert. It might fall in a rain forest. But wherever it fell, the mapmaker would come away with a wildly inaccurate picture of our world.

Geller found a better strategy. The confetti strip that she cuts from the map of the world is so thin that its area adds up to no more than that of Rhode Island. But because it is so long, it contains a remarkable amount of information. Chosen randomly, this particular slice happens to begin in northern Russia and then angles southeast, passing over Japan before arcing across the Pacific Ocean. Thus far, it has revealed the existence of a land mass and a body of water, both quite large. It has also given evidence of a smaller land mass—Japan—and several brief stretches of water corresponding to rivers. As it proceeds further across the globe, the slice will cross the Andes, revealing the existence of mountains, and will include arctic latitudes (snow, glaciers) as well as equatorial ones (tropical islands). Much of what the ribbon shows will be uninterpretable, of course—is that tiny bit of water a pond or a river? are mountains isolated peaks or do they form ranges?—but the information obtained by this strategy is far greater than had the region surveyed been a patch.

The three-dimensional analog of Geller's confetti strip is a fan-shaped slice of the universe; a survey of the redshift of every galaxy lying in a thin, arcing path across the sky. By the time John began gathering the data, preliminary evidence of large-scale structures in the universe had been found. On the other hand, neither he nor Margaret found this evidence compelling. Both of them implicitly accepted the notion of the essential uniformity of the universe, and neither thought their survey was going to find anything particularly unusual. For this reason, neither Huchra nor Geller encouraged de Lapparent to plot up the data as it came in. Not until the survey was complete did anyone realize what they had accomplished.

The pattern they found might have been discovered by Davis, had his survey been designed differently. Rather than studying a patch of the universe, the most obvious strategy, or a slice à la Geller and Huchra, Davis and his group had taken redshifts of the brightest galaxies in a wide region of the sky. Their strategy had been analogous to surveying various isolated backyards randomly peppered across a continent. The group, in fact, measured the redshifts of many more galaxies than Geller and Huchra. But because their data was so dilute, it showed little evidence of structure. Indeed, one of the figures in the paper they published reporting their work happens to cover the very region of the Geller-Huchra slice. But so sparse is the coverage that even with the benefit of hindsight, it is impossible to make out in that figure the characteristic voids and arcing sheets.

These patterns strain our understanding. They differ radically from what we had originally expected. Gravitation, the attraction of each galaxy for every other, is widely believed to be the only force acting to determine the structure of the universe as a whole. But gravitation would draw galaxies into clusters, each roughly spherical, much as it draws together all the stones, clods of dirt, and drops of water that comprise the Earth into a globe. And although clusters do exist in the universe, the foam-like appearance of the Geller-Huchra map bears little resemblance to this expected pattern.

Huchra is bearded, irreverent, and possessed of a sharp intelligence that often manifests itself as wit. It is with an amused and ironical eye that he regards the world's follies. Familiar with theoretical issues, John is nevertheless strongly distrustful of anything he regards as an excessive flight of fancy. He takes pleasure in poking holes in the beautiful dreams of the theoreticians. This he accomplishes by referring to his encyclopedic knowledge of hard, prosaic facts.

Even for an astronomer, his personal acquaintance with the universe is remarkable. The sky is his backyard. Huchra can tell you the brightness of the third star from the end of the handle of the Big Dipper. He can sketch on a pad the shape of any galaxy you might care to name. He likes facts, in and of themselves. It is characteristic that in our conversations he remembered his weight as a freshman in high school, and the time it took him to wrestle an opponent to the mat. He also likes lists of facts, and the sense of closure to be gained by completing one. His approach to his hobby of mountaineering is that of the "peak bagger," who keeps lists of the summits he has reached— one-third of the big mountains of Colorado so far, and all but one of the San Gabriels outside of Los Angeles. It is the same mentality that, nighttimes in a dome on Mount Hopkins, works through another kind of list. John told me one day that he has personally measured the redshifts of more galaxies than any other astronomer, and quite possibly more than the sum total of all the other astronomers who have ever lived.

In the midst of the survey not long ago, he stumbled upon the closest instance of a gravitational lens, in which light from a distant quasar is being focused toward us by an intervening galaxy. He resists the term "luck" to describe this discovery. "I have no use for the notion that an astronomer is a person who simply points a telescope at some random piece of the sky and takes a picture. To do science, you must be engaged in a serious project, with a well-defined goal—and then you are likely to find something other than what you set out to find."

All scientific fields evolve, he feels, and they evolve primarily by the invention of new techniques. "Infrared astronomy, for example, was created primarily by solid state physicists who had developed all these marvelous instruments that responded to infrared radiation, and who then looked around for something to do with them. These people will probably go on thinking of themselves as solid state physicists to the end of their lives—but their students will think of themselves as infrared astronomers, and *their* students will simply think of themselves as astronomers. The new technique will have become amalgamated into the fabric of the profession. It will have become invisible.

"But this is not my style of science. I would argue—and I recognize that not all people feel this way, and I would not even argue that everyone ought to feel this way—that it is vitally important to be driven not by techniques, but by questions, and in the pursuit of these questions to use every technique at one's disposal." Huchra is a generalist ("Damn few of us left"), and he has personally made observations with telescopes adapted to every region of the electromagnetic spectrum, from long-wavelength radio telescopes, to optical telescopes, to an X-ray telescope orbiting the Earth. I know of few astronomers whose breadth of expertise comes close to his.

Because he is widely recognized as being so technically accomplished an observer, and because he is such a generalist, people working on differing problems approach Huchra from time to time with requests for help. Thus he gets into new areas—at any given moment, he might be working on fully twenty different projects. They range in size. On one occasion, he dashed cold water on a theoretical flight of fancy with an observation performed in the space of a mere few hours. At the other end of the range is his redshift survey work, which occupies half his telescope time and, as he puts it, "pays the rent."

In the summer of 1990 the MacArthur Foundation telephoned Geller with the news that she had received one of their "genius" grants. Her reaction was to ask whether Huchra was getting one, too.

Geller and Huchra have collaborated for essentially all of their careers. Their scientific styles are perfectly complementary: Huchra the virtuoso of the telescope, deeply familiar with the sky; Geller the interpreter, with interests ranging far beyond the boundaries of the profession. Their personal styles are complementary as well. "I have no patience," says Margaret with a broad smile. "And John has lots. That's how we get these big projects done—because I have no patience, I drive them forward, and because he does, he makes sure we actually complete the work." At one point John mentioned that he was a highly organized person: when I asked if he differed from Margaret in this regard, he burst out laughing.

His laugh was one of affection and respect. Although there is no romantic love between them, the two are very close friends. They speak often of personal matters, and each is available to the other if needed for comfort and support. For all her delight at the MacArthur grant, the news that she had received it came as a mixed blessing to Margaret, for John had not. It helped that soon thereafter they won a second award together.

Perhaps one source of this mutual affection lies in a deep similarity between the two. Both grew up thinking of themselves as outsiders. In my conversations with them, I was continually struck by the discrepancy between their accounts of their pasts and the people I saw before me. One of Huchra's most vivid recollections of his childhood was of his small stature—a shrimp, and the butt of bullies. But he is now a person of perfectly respectable size. He referred often to me to the sense of alienation he felt in college, and which has not entirely left him even now: in many ways, he feels closer to people like his parents, people who care not the slightest about scientific research, than to his own colleagues. But among these colleagues he has reached a position of eminence.

Margaret describes herself as a shy child. She is not shy anymore—not by a long shot. Recently, Radcliffe organized a program including lectures by practicing scientists for young women of high school age interested in science, in an attempt to do something about the lack of women in the field. Margaret heartily approved of the project, until she learned that only women would be allowed to address the program's shrinking violets. She promptly fired off a letter, and then boomed into the office of the college's president to let her know, in no uncertain terms, what she thought of the restriction.

Geller is a person of strong opinions, strongly expressed. Her conversation zooms rapidly from topic to topic, and it is forceful and confident—she fills the room. She has immense energy, and seldom seems to slow down. She is disarmingly open, and speaks frankly about herself as a person. Somehow, Margaret manages to speak of her great professional success with neither undue modesty nor braggadocio. Remarkably for a person of such strength, she is also a person of great friendliness and warmth.

When Margaret was a child she kept a stamp collection, and she recalls thinking that Italy must have been an ugly country because its stamps were so unattractive. She is, in fact, a highly visual person, and some of her most intense childhood memories are of paintings. "It is amazing how the images that I saw as a child have stuck in my mind. I don't happen to own art books with those particular images, but I *have* them in my mind. They have stayed with me in extraordinary ways." Conversely, she does not have equally strong childhood memories of books, nor of scientific concepts. The professional style that Geller ultimately evolved turned out to be intensely visual as well—witness the essentially geometrical nature of her strategy for mapping the universe.

Down the hall from John's and Margaret's offices is a door engraved with the names Immanuel Kant and Fritz Zwicky (Zwicky was an early cosmologist). These are two desktop computers. "We loaded our redshift data into them," she said, "and we developed an image processing system which would display the map we had made. I remember sitting there and just staring at it, as if it were some miracle that had come to pass, watching for hours as the computer would rotate the image this way and that, and just being fixated upon it. And it didn't take us long to realize that we couldn't be the only people who would be captivated by these natural patterns."

Geller has been instrumental in bringing their map of the universe before the general public. She lectures incessantly. In collaboration with a filmmaker, she has produced a video describing the redshift survey. When I visited her, they were at work on a second film, which includes a computer-animated spaceship voyage through the structures they have discovered. It is an extraordinary film: I found myself actually growing dizzy as we zoomed along the Great Wall at velocities far exceeding that of light.

Among all the images of modern science, their slice of the universe with that strange, primeval stick figure (page 118) has struck a responsive chord. It is a uniquely evocative image. Margaret estimates that it has been seen by perhaps 100 million people worldwide. Ironically, had they elected to survey any other slice, the stick figure would have been absent, for it is a localized structure. Maps of other regions of the universe contain just as much interesting scientific data, but none carries the same impact. I often catch myself wondering whether the reaction to their work would have been so great, even among scientists who pretend to be immune to such things, had they happened to map a different region.

"People sometimes ask me," says Geller, "how we can justify spending so much money on science rather than, for instance, the homeless. My response is that this is not the choice. This is a high-tech investment, and the choice is between supporting science and supporting something more like weapons research. You have to be careful that your expenditures produce a climate in which people can be imaginative. If society doesn't invest in an art, which I think science is, then society dies. This business of mapping the universe is part of our attempt to write our address on a very grand scale. By studying this ancient light, we are reaching for an understanding of how the universe is made."

The attempt to write our address in the universe is the attempt to speak about the universe *as a whole*. But what we say about the cosmos depends on where we live. A primitive South Sea islander might be tempted to say the world consists solely of water. A Bedouin nomad would argue that it consists of nothing but sand.

With the enlargement of one's view comes the recognition that these are not descriptions of the universe as a whole. They are merely descriptions of local conditions. Every advance in the technology of exploration has allowed us to draw maps on larger and larger scales—from the immediate environs of some particular city-state to the entire Mediterranean Sea, from Europe to the Earth as a globe. Expanding the scope still farther has revealed the Earth orbiting the Sun, surrounded by a more or less ill-defined sphere of stars. This is the view put forth by the scientific revolution. Yet larger views have probed the distribution of stars in space.

And with every advance, our opinion of what the universe as a whole consists of has changed—and it has changed completely. So how broad a view do we need in order to terminate this perpetually evolving process? How can we pass beyond discussing mere local conditions, and reach an understanding of the universe itself?

The problem is that the universe may well go on forever. It may be infinite—and no matter how gargantuan the telescopes we construct, no matter how enormous the regions of space we survey, our view of an infinite universe is guaranteed to be hopelessly parochial. Such a cosmos cannot possibly be explored in its entirety. If the universe actually is infinite, we will never be able to fathom all its mysteries.

Indeed, to entertain the notion that the universe is *not* infinite is to become embroiled in paradox. For if the cosmos is limited, and has an edge, what is to prevent us from going out to that edge—and then one step further? Einstein found a way around this conundrum with his theory of curved space, within which the universe, like the surface of the curved Earth, is finite and yet possesses no edge. But this is only one of several models of the cosmos proposed by Einstein; the rest, including the one currently most popular among cosmologists, are infinite. The problem seems inescapable: no understanding of the universe as a whole may ever be possible.

The early decades of this century witnessed what appeared to be a wonderful solution to this problem. Evidence was found that the distribution of galaxies might be uniform. And if the universe was uniform, each part would be representative of the whole. The most minute fraction of a homogeneous cosmos tells us everything there is to know about it. In this case, what we learned from our limited, hopelessly parochial surveys would actually be of universal significance. The apparent uniformity of the universe allowed us to penetrate beyond the parochial, and to attain a glimpse of the ultimate. It made the science of cosmology possible.

A heady discovery. But Geller and Huchra's survey has shown this "discovery" to be false. They have shown that the apparent uniformity of the distribution of galaxies was no more than an illusion. It was a mistake, to which we were led by the absence of distance indicators in the universe. It is always possible, of course, that their foamlike structure, when surveyed over yet more gigantic scales, will itself turn out to be uniform. In this case the postulate of the uniformity of the universe can be retained. But the lesson of history gives little reason for hope in this regard. Surely in the past, with each enlargement of scale, our notion of what the universe consists of has changed utterly.

The world as it presents itself to us in daily experience is of overwhelming complexity. The dream of science is that beneath this complexity is an underlying simplicity. The life of every scientist is dedicated to the hope that a viewpoint can be found in which the world will turn out to be simple. But everyone recognizes this principle for what it is—a hope. No scientist worth his or her salt ever claimed the universe *had* to be uniform. The postulate was the expression of a dream.

Geller and Huchra's discovery has triggered a burst of work on the problem of large-scale structure. By now their group has completed surveys of a great many more slices, working layer by layer toward a full three-dimensional map of our corner of the universe. Numerous other groups have also gotten into the field. Some have surveyed slices; others, focusing attention on individual patches of the sky, have pushed to enormous depths, yielding pencil-thin "core samples" drilled through the foam. Hints of yet more gigantic structures have been reported.

"Because of the success of the redshift survey," says Geller, "we have upped the ante. The field has become very hot, and the competition is severe. The various projects are getting even larger; we are asking people for quite large commitments. Instruments are being designed which will cost millions of dollars, and which will not go on-line for years. John and I now run a group of twenty people. It's very flattering, but at the same time it takes enormous chutzpah, and things can be quite scary sometimes."

In addition to science, Margaret finds herself nowadays thinking more and more about the visual arts. She spent several days with the photographer Berenice Abbott shortly before she died, and she is turning over in her mind plans for a film on Abbott's life and work. As other ideas for films occur to her, Margaret jots them down in her calendar: page after page fill up with notations in her neat, precise hand. She finds that the MacArthur Award opens doors for her as she moves into this new field. At the same time, however, the freedom the award has brought carries with it an almost intimidating sense of responsibility for her future.

As a result of their success, John and Margaret find themselves increasingly caught up in administrative duties. Huchra serves on an ever increasing number of professional and government committees, and has been named associate director of the Center for Astrophysics. Geller spends several days each week on administration and fundraising for their group's work. Weeks pass in which they do not see each other.

Sometimes, late in the evening when Margaret is at home, the phone will ring. It is John, calling from the mountaintop in Arizona. As stars shine in through the slit in the dome overhead, as he waits for the Z-machine to gather its data, the two will talk. They will talk for hours—of ideas for new projects, of science, of friends. Occasionally there will be a pause, as John swings the telescope to observe a new target.

Sources

1. The Ladies of Observatory Hill

"He was rude to many..." P. Kidwell, "Cecilia Payne-Gaposchkin" in *Uneasy Careers and Intimate Lives: Women in Science 1789–1979*, eds. P. G. Abir-Am and D. Outram (New Brunswick, N.J.: Rutgers University Press, 1987). All other quotes are from K. Haramundanis, ed., *Cecilia Payne-Gaposchkin* (Cambridge, England: Cambridge University Press, 1984).

2. The Bulldog

M. J. Klein's discussion can be found in E. G. D. Cohen and W. Thirring, eds., *The Boltzmann Equation* (Vienna: Springer-Verlag, 1973).

3. The Magician

All quotations are from George Gamow, *My World Line* (New York: Viking Press, 1970).

4. A Gentleman of the Old School

"I felt that the physical conditions ...," V. S. Naipaul, *India—A Million Mutinies Now* (New York: Viking Press, 1991) (c 1990). "A man in the course of ...," Homi Bhabha's inaugural address before the United Nations Conference on the Peaceful Uses of Atomic Energy, 1955. "The acquisition by man ...," *ibid.*

5. Luie's Gadgets

The two quotations from Oppenheimer's security hearing are from "In The Matter of J. Robert Oppenheimer: Transcript of Hearing Before Personnel Security Board, Washington D.C., April 12, 1954 through May 6, 1954" (Washington, D.C.: Government Printing Office, 1954). All other quotes are from Luis Alvarez, *Alvarez* (New York: Basic Books, 1987).

6. All Genius and All Buffoon

"I was inspired . . . ," R. P. Feynman, "The Development of the Space-Time View of Quantum Electrodynamics," Nobel lecture, in Les Prix Nobel en 1965, Nobel Foundation, Stockholm (1966). "You can know the name . . . ," quoted in Jagdish Mehra, *The Beat of a Different Drum* (New York: Oxford University Press, 1994), 4. "It was obvious . . . ," quoted in Mehra, *op. cit.*, 229. "I'll prepare a freshman lecture . . . ," David Goodstein, "Richard P. Feynman, Teacher," in *Most of the Good Stuff*, eds. Laurie Brown and John Rigden (New York: AIP Press, 1993), 122. "After working some more . . . ," Richard P. Feynman, *Surely You're Joking, Mr. Feynman!* (New York: W. W. Norton, 1985), 252. "There are two kinds of geniuses . . . ," Mark Kac, *Enigmas of Chance* (New York: Harper & Row, 1985), xxv. "For a long line there are . . . ," R. P. Feynman, "Application of Quantum Mechanics to Liquid Helium," in *Progress in Low Temperature Physics*, vol. 1 (New York: Interscience Publishers, 1955), 52 (I have slightly simplified some of the mathematics here).

7. Big Science

"Looked out over a sea . . . ," Daniel Kevles, *The Physicists* (Cambridge: Harvard University Press, 1987). "A secret army . . . ," Gerald Holton, "Scientific Research and Scholarship: Notes Towards the Design of Proper Scales," *Deadalus* (Spring 1962). "Had been waiting . . . ," Gary Taubes, *Nobel Dreams* (New York: Random House, 1986). "Resembling some vast . . . ," Report of the SSC design study group, quoted in *Physics Today* (Sept. 1989), 53. "The universe astonishes . . . ," Sheldon Glashow and Leon Lederman, "The SSC: A Machine For the Nineties," *Physics Today* (Mar. 1985), 28. "This quotation describes . . . ," Martin Perl, *Tau Physics* (Stanford, CA: SLAC Publications, 1993).

8. *Our Address in the Universe*
"There is a long history . . . ," *Science* 255, (13 Mar. 1992), 1383.
"I have light brown . . . ," *Science* 263, (11 Mar. 1994), 1480.

Acknowledgments

Much of the material contained in these profiles was gleaned from interviews—interviews sometimes conducted with my subjects, sometimes with their colleagues, and always with people who were experts on their careers or on their work. These people gave generously and unstintingly of their time, and I am deeply grateful to them: the book could never have been written without their help. Thanks to

Chapter 1: Peggy Kidwell, Patrick Williamson, and my parents, Jesse and Naomi Greenstein.

Chapter 2: Jacqueline Waldman, for her translation of Boltzmann's account of his trip to America.

Chapter 3: David Inglis and Patrick Williamson.

Chapter 4: Particular thanks to Kannan Jagannathan for much assistance; thanks also to S. Chandrasekhar, J. J. Bhabha, R. M. Lala, J. Maddox, D. K. Malegamvala, N. Mott, R. Narasimhan, D. Shoenberg, O. Siddiqi, V. Singh, B. V. Sreekantan and B. M. Udgaonkar. The travel that made this project possible was financed by a grant under the Amherst College Research Awards Program.

Chapter 5: Walter Alvarez, Philip Morrison, and Jack Sandweiss.

Chapter 6: Don Coyne.

Chapter 7: Michael Kreisler and Martin Perl.

Chapter 8: Margaret Geller, John Huchra, and Alan Sandage.

Index

Numbers in *italics* indicate pages with photographs or illustrations.